DATE DUE

GAYLORD	PRINTED IN U.S.A.

T

Shiphandling Simulation
Application to Waterway Design

Committee on Assessment of Shiphandling Simulation
Marine Board
Commission on Engineering and Technical Systems
National Research Council

William C. Webster, Editor

National Academy Press
Washington, D.C.   1992

The program described in this report is supported by Cooperative Agreement No. 14-35-0001-30475 between the Minerals Management Service of the U.S. Department of the Interior and the National Academy of Sciences.

Library of Congress Catalog Card Number 92-60999
International Standard Book Number 0-309-04338-7

DEDICATION

Mr. C. Lincoln Crane, Jr., a world renowned expert in ship maneuverability, directed this assessment until his death in September 1989. His contributions to the committee and ship maneuvering research were substantial. His sudden, untimely and heroic death was an event that touched each committee member deeply. We have lost a good friend; naval architecture has lost a respected leader.

C. Lincoln Crane, Jr., was posthumously awarded the Gold Lifesaving Medal by the Department of Transportation for his rescue of a woman swept out to sea by dangerous surf conditions.

COMMITTEE ON ASSESSMENT
OF SHIPHANDLING SIMULATION

Committee Members

WILLIAM C. WEBSTER, *Chairman*, University of California at Berkeley
WILLIAM A. ARATA, Biscayne Bay Pilots, Miami, Florida
RODERICK A. BARR, Hydronautics Research, Fulton, Maryland
PAUL CHILCOTE, Port of Tacoma, Washington
MICHAEL DENNY, ShipSim Corporation, Northport, New York
FRANCIS X. NICASTRO, Exxon Company International, Florham Park, New Jersey
NILS H. NORRBIN, SSPA Maritime Consulting AB, Gothenburg, Sweden
JOSEPH J. PUGLISI, U.S. Merchant Marine Academy, Kings Point, New York
LEONARD E. VAN HOUTEN, L. E. Van Houten and Associates, Greenwich, Connecticut
JAMES H. VINCENT, Systems Control Technology, Palo Alto, California

Government Liaisons

H. PAUL COJEEN, U.S. Coast Guard
LARRY L. DAGGETT, U.S. Army Corps of Engineers, Waterways Experiment Station
FREDERICK SEIBOLD, U.S. Maritime Administration
DAVID A. WALDEN, Carderock Division, Naval Surface Warfare Center

Marine Board Staff

CHARLES A. BOOKMAN, Director
C. LINCOLN CRANE, JR., Project Officer (until September 1989)
WAYNE YOUNG, Project Officer (from September 1989)
CARLA D. MOORE, Project Assistant
JUDITH GRUMSTRUP-SCOTT, Editorial Consultant

iv

v

Preface

BACKGROUND

The nation's ports and waterways are vital links in national, regional, and local intermodal transportation and economic systems. The safety of vessel operations in these waters and ultimately the underlying waterway design are under increasing scrutiny as a result of major shipping disasters on all coasts. At the same time, the overall costs of waterway projects, increased cost-sharing responsibilities of local project sponsors, and awareness of environmental impacts has increased pressure for more efficient waterway designs. This pressure in turn has motivated new and improved techniques to offset the traditional approach to waterway design, an approach that can result in channels of questionable safety, excessive cost, or both because of uncertainty, conservatism, and reliance on rules of thumb.

The economic importance of the ports and waterways system is reflected in the flow of cargoes through the system and in the substantial national investment to support waterborne commerce. About one-third of domestic intercity trade and almost all foreign trade by weight pass through the system each year. The annual waterway investment by the U.S. Army Corps of Engineers (USACE) alone is $1.23 billion. Additional federal government investments include construction, operation, and maintenance of aids to navigation.

State, port authority, and commercial investment has until recently focused principally on port facilities, including berthing and cargo-handling capabilities. Funding of maintenance and modernization projects to assure efficient operation of the waterway system was primarily a federal responsibility until passage of the Water Resources Development Act of 1986. The act shifted much of the financial responsibility for expensive waterway improvement projects to local sponsors. This fundamental change in policy and escalating costs increased the importance of most-cost-effective waterway design as a means to help keep construction and maintenance costs affordable.

Shiphandling simulator technology is considered by many design engineers to be a potentially important and effective tool for waterway design. Interest is growing in the use of simulation technology to increase confidence in waterway designs and reduce the costs of construction and maintenance.

Shiphandling simulations based on available technology have been developed over the last 3 decades. Simulations have been used in a variety of contexts from training vessel crews and analyzing marine casualties to the evaluation of buoy placement. The use of simulation in the design process for modifying or developing channels and waterways is the most technologically demanding of these applications. Confidence in the application of simulators in channel design has been hampered by difficulties in assuring that the results of simulations reproduce what would have occurred in the real situation or provide sufficient value to justify their expense.

The related issues of choosing a simulator facility with suitable capabilities to address the design problem effectively, of having confidence in the results, and of integrating simulation results into the waterway design process were studied by an interagency committee in 1986. That study, convened and coordinated by the U.S. Army Corps of Engineers, recommended consultation with the National Research Council (NRC) on the role of shiphandling simulation in waterway design and supporting research (USACE, 1986b).

NRC STUDY

The NRC convened the Committee on Assessment of Shiphandling Simulation under the auspices of the Marine Board of the Commission on Engineering and Technical Systems.

Committee members were selected for their expertise and to ensure a wide range of experience and viewpoints. The principle guiding the constitution of the committee and its work, consistent with the policy of the NRC, was not to exclude members with potential biases that might accompany expertise vital to the study, but to seek balance and fair treatment. Committee members were selected for their experience in port and waterway design, hydrodynamic and mathematical modeling, computer simulation, statistical analysis, ship control design, aviation and shiphandling simulation technology, and shiphandling. Academic, industrial, government, and international perspectives were also reflected in the committee's composition. Biographies of committee members are provided in Appendix A.

The committee was assisted by the U.S. Army Corps of Engineers, U.S. Coast Guard, U.S. Maritime Administration, and the U.S. Navy's Carderock Division, Naval Surface Warfare Center (formerly David Taylor Naval Ship Research and Development Center), all of which designated liaison representatives.

The committee was asked to conduct an interdisciplinary assessment of the state of practice of simulation of ship transits in restricted waterways, the adequacy of data input to simulators, and the validity of hydrodynamic and related models. The committee was further asked to develop guidance for determining the applicability and presentation of simulation results, provide guidance for determining the required and achievable accuracy of simulator results, and recommend research to resolve any discrepancies. However, assessment of human factors in shiphandling simulations, shiphandling theory, and waterway design theory were beyond the scope of study. The issue of competitive advantage associated with the economic potential of port regions to sponsor waterway projects, although an important factor in assessing the effects of the Water Resources Development Act of 1986, was also beyond the scope of study.

The committee reviewed available data and literature to determine the state of practice of simulator use in maritime activities, including the appropriateness of various levels of simulation to different port and waterway design objectives. This examination was supplemented by visits to simulator facilities and discussions with experts in the United States, Europe, and Japan, which were documented in detailed trip reports. Case studies of shiphandling simulator application to waterway design were developed and are included as Appendix C. A source reference list on mathematical models was prepared and included as Appendix D.

REPORT ORGANIZATION

The audience for which this report was prepared consists of waterway designers, naval architects interested in the scientific issues involved with predicting the forces acting on a ship as it maneuvers in a constrained waterway, simulation experts knowledgeable in the computational and graphical presentation aspects of the technique, members of the maritime and general public who participate in the waterway design process, and decision makers affecting the use of simulation. Understanding shiphandling simulation for waterway design requires a simultaneous understanding of the science and practice of simulation and the context of waterway design in which simulation is used. Chapters 2, 3, and 4 provide this information as background because suitable, concise references are not available.

Chapter 1 provides an overview of the relationship of U.S. ports to the economy, port modernization needs, trends and issues affecting port and waterway development, general goals of waterway modernization, and how shiphandling simulators are used in achieving these goals.

Chapter 2 summarizes the harbor and waterway design process. It identifies the process used, participants, and design factors and issues.

Chapter 3 describes shiphandling simulators and their use in the waterway design process.

Chapter 4 discusses the two principal types of shiphandling simulations, those operating in real-time mode with human operators in the decision-making loop and those operating in fast time with human operators replaced by computer-based pilot models.

Chapter 5 discusses and assesses mathematical models used in channel design simulation.

Chapter 6 assesses simulator technology and the validity of using this technology in the design process.

Chapter 7 discusses practical applications of simulators in harbor and waterway designs.

Chapter 8 identifies research needs, including the framework for analysis and results, mathematical models, simulator fidelity, and guidelines for the level of simulation.

Chapter 9 provides the committee's conclusions regarding the state of practice and recommendations for using simulators in the waterway design process.

Acknowledgments

The committee gratefully acknowledges the generous contributions of time and information provided by liaison representatives, their agencies and organizations, shiphandling simulation practitioners, and the many individuals in government and other organizations interested in the application of shiphandling simulation to channel design.

Larry L. Daggett, U.S. Army Corps of Engineers, Waterways Experiment Station, participated in the site visits and in-depth interviews held in Europe, and provided technical support and reference materials. H. Paul Cojeen, U.S. Coast Guard, provided technical advice on navigational safety factors in design. Frederick Seibold, U.S. Maritime Administration, provided technical advice on his agency's prior research using simulators. David A. Walden, Carderock Division, Naval Surface Warfare Center, provided technical advice on ship hydrodynamics.

Special thanks are extended to members of the international design and shiphandling simulation community who met with the committee's delegation during its European visit and provided technical advice on the state of practice. The committee is indebted to: S. D. Sharma, Institute für Schiffbau, Hamburg; T. E. Shellin, Germanischer Lloyd, Hamburg; A. H. Nielsen, M. S. Chislett, and L. Wagner-Smitt, Danish Maritime Institute, Lyngby; M. Oosterveld, V. ten Hove, Jan P. Hooft, and J. Perdok, Maritime Research Institute, Wageningen, The Netherlands; W. Veldhuyzen, I. Onassis, Ing. W. de Joode, The Netherlands Organization for Applied Scientific search, Delft, The Netherlands; J. W. Koeman and W. Ph. van Maanen, Port of Rotterdam; M. P. Bogarerts, Rijkswaterstaat, Rotterdam; and Ian McCallum, Maritime Dynamics, Lantrisant, Wales.

The committee gratefully acknowledges the many contributions of the late C. Lincoln Crane, Jr. As staff study director, Mr. Crane coordinated the committee's activities including the European trip, provided expert technical advice on ship maneuvering, and supported development of background papers and case studies.

The extraordinary cooperation and interest in the committee's work of so many knowledgeable individuals were both gratifying and essential.

Contents

Executive Summary

Ports and waterways are vital links in local, regional, and national intermodal transportation and economic systems. The safety of vessel operations in these waterways and ultimately the underlying design are under increased scrutiny as a result of major shipping disasters on all coasts. Related issues are the standard shipping practice of scheduling large ships into waterways originally designed for smaller, earlier-generation vessels, the lengthy time needed for design through construction of waterway modernization projects, and the lack of impetus for design reevaluation for safety after a waterway has been constructed or altered.

Traditional waterway design practice relies heavily on rules of thumb and conservatism for margins of safety. At the same time, the overall costs of waterway projects and expanded cost-sharing responsibilities of local project sponsors imposed by the Water Resources Development Act of 1986 have increased pressure for more cost-effective waterway designs. One effect of the revised cost-sharing responsibilities has been to stimulate efforts to develop design tools that improve the cost-effectiveness of design.

Over the past several decades, development of some waterway designs in the United States and overseas has been aided by the use of hydrodynamic physical scale model and computer-based shiphandling simulations. Each provides alternative means for achieving refinements in design not verifiable with other design tools. New attention has been focused on the potential of simulations to improve cost-effectiveness while still providing adequate margins of safety. These two approaches provide different capabilities and levels of control for assessing alternate project dimensions relative to ship behavior under the influence of human operators and environmental conditions.

Interest has increased in the use of computer-based simulations. With this medium, projects can be modeled mathematically rather than physically, conceptually permitting the modeling of any waterway with the same hardware; vessel pilots can be presented with realistic representations of the operating environment under varying but controlled conditions; and the computer capability for high-speed, automated simulations using mathematical models of pilot behavior can be used to generate a large number of vessel transits that would not be feasible in real time. The use of simulation for these purposes, although promising and used in several major waterway studies, has been incorporated in only a small number of waterway projects.

This study addresses three questions about the use of computer-based simulations for waterway design:

- Does simulation work?
- When should simulation be used?
- How can simulation be enhanced as a design aid?

DOES SIMULATION WORK?

Computer-based shiphandling simulations sponsored by government, port authorities, and the maritime industry have been used effectively as a waterway design tool by planners and engineers. The technique provides an improved means to assess the operability of a proposed waterway improvement by approximating vessel behavior in the full waterway operating environment, thereby offsetting the traditional reliance on rules of thumb to provide adequate margins of safety.

Six applications of simulation to channel design were selected for detailed examination by the committee after review of over 50 different applications for which detailed results were available. The six simulation studies chosen for case study included a wide range of situations and are representative of typical applications that could be applied to design studies for U.S. waterways. The six simulations examined were:

- a study sponsored by the Exxon Corporation (1980-1981) to determine the maximum-size oil tanker that could safely transit a narrow channel cutting obliquely across the Coatzacoalcos River, Mexico, at the entrance to a tanker loading facility;

- a State of Virginia-sponsored study (1980–1986) to improve existing channel designs for Hampton Roads ports so as to permit safe transit of deep-draft coal colliers in channels with 55-foot depths.

- A U.S. Army Corps of Engineers (USACE) study (1983–1984) performed by the agency's Waterways Experiment Station to verify the validity of the final design for a major ship channel improvement project in Richmond, California, that would permit the discharge of fully loaded 85,000-deadweight ton (DWT) tankers and partially loaded 150,000-DWT tankers.

- A Panama Canal Commission study (1983–1986) to determine the specific dimensions of the optimum navigation channel that would afford a reasonable balance between excavation cost and safety of modifications necessary to permit two-way traffic of Panamax-size vessels throughout the canal's length.

- A study sponsored by Port of Grays Harbor, Washington, (1986) and performed by the USACE Waterways Experiment Station to verify the feasibility of the final design for widening and deepening 24 miles of an estuary and bar channel, improving a highway bridge fender system, and replacing a railroad bridge.

- A study sponsored by Port of Oakland, California, (1986–1988) to develop alternative channel designs for the inner and outer Oakland harbors in order to find suitable designs that would open the port to larger, more cost-efficient containerships.

Scientific, quantitative validation of the results of simulations is not yet available. However, pilot participation in the validation process and pilot acceptance of simulations indicate that reasonable success can be achieved with the existing state of practice by re-creating a realistic piloting experience through modeling of waterway complexities, the physical environment, and operational factors. The case studies revealed that simulations can effectively aid in decision making by providing unique quantitative information for answering design questions associated with channel depth, width, geometry, dredging requirements, aids to navigation requirements, and tugboat assistance. Additionally, simulations have also provided a unique, common forum for discussion between design participants and an easily understood context for problem identification, conflict resolution, and decision making.

In some of the applications examined, the construction cost savings stemming from design changes developed using simulation were much greater than the cost of simulation. The committee believes that risks to shipping and the environment can be reduced through design refinements based on simulations, but this reduction is difficult to assess or express in monetary terms. Nevertheless, evidence from the six case studies shows that simulation technology can be effectively applied to the waterway design process with substantial benefits. Simulation models developed for waterway design can also be used in simulations conducted for training.

WHEN SHOULD SIMULATION BE USED?

Simulation should be used when:

• *Vessel operational risk is a significant design issue.* Incorporation of human pilot skills and reactions in the prediction of the behavior of a vessel in a proposed waterway is unique to shiphandling simulation. Differences in risk resulting from a variety of critical environmental conditions can be identified. Aids to navigation requirements that can further reduce risk can also be assessed.

• *Cost and design optimization is an issue.* The effect on risk resulting from variations in the many design factors that define a waterway can be evaluated. This capability is an important decision aid in the assessment of the components of life-cycle costs. Simulation is particularly useful for assessing operational differences between design alternatives.

• *Competing interests among technical and nontechnical participants in the waterway design process are an issue.* Simulation provides a unique way to bring critical and contentious aspects of the design into focus. Design modifications to accommodate competing interests can be tested and the consequences displayed in formats that do not require technical expertise to assimilate and understand.

Because elements of these three issues are frequently associated with most waterway designs, the committee concluded that shiphandling simulation should be developed as a standard tool available for use in waterway design. The level of sophistication of simulations needed for this process depends on the particular design. However, guidelines for the appropriate level for a given situation are not available within the current state of practice.

HOW CAN SIMULATION BE ENHANCED AS A DESIGN AID?

Simulation is a highly technical art involving the integration of many skills: naval architecture, civil and marine engineering, piloting, computer techniques, and human engineering. In all of these areas, there are substantial unresolved issues. Confidence in the use of shiphandling simulation for waterway design is limited by issues of fidelity and the level of simulation required. Use is inhibited by cost, scheduling, and interpretation of the results. More use of simulation in the waterway design process could be motivated by:

• Reducing the costs of simulation.

• Developing a definitive guide to assist designers in choosing a simulator for specific applications. Although the cost of computer equipment needed for simulation has dropped significantly in recent years, the cost of the labor-intensive set up and conduct of the simulation has increased. These latter costs and the duration of the simulation process are sensitive to the level of simulation required, but no guidelines exist for this choice.

• Developing minimum requirements for fidelity and validation of mathematical models of ship dynamics, waterway data bases, and the simulator environment including visual displays and bridge mock-up.

• Developing a better understanding of the behavior of ships in situations unique to waterway design. These situations include operation in the following conditions: with small under-keel clearance, near banks of arbitrary geometry or muddy bottoms, in sheared currents, and in close passage of other ships.

• Developing and validating a mathematical framework for extrapolating the results from a small sample of simulation runs to a prediction of the performance of future traffic in the waterway.

• Establishing a carefully composed, interdisciplinary validation team as a formal element in each simulation validation process.

SUGGESTED RESEARCH

Confidence in simulations can be increased through a systematic research program designed to address the preceding deficiencies. The committee recommends implementation of a research program that

• assesses the need for fidelity in the mathematical models and simulator hardware,

• develops ways to determine, assess and resolve the uncertain elements in mathematic models, and

• provides a capability for interpreting the results.

The research program should be coordinated by the Army Corps of Engineers in cooperation with other interested federal agencies and segments of the maritime community and in consultation with organizations representing the best technical expertise available within the waterway design and simulation community.

Shiphandling
Simulation

1

Introduction

WATERWAY MODERNIZATION

The nation's ports and 28,000 miles of navigable waterways authorized for improvement under federal programs are vital to national, regional, and local transportation and economies. They provide a critical intermodal link to the global economy while also serving as local and regional employment centers. Over 30 percent of the nation's domestic intercity freight trade and 99 percent of overseas trade by weight (74 percent by volume) pass through this system as waterborne commerce. Although variable from year to year, the flow of cargoes through U.S. ports hit 2.09 billion tons in 1988 (U.S. Maritime Administration, 1990). However, existing ports and waterways do not adequately accommodate the most modern ships in terms of efficiency, safety, and cargo handling capabilities. Thus, there is interest nationwide for modernization of ports and waterways systems to accommodate modern ships and maintain competitive advantage in regional and international trades (Frankel, 1989; *Journal of Commerce*, 1991a; Kagan, 1990; U.S. Maritime Administration, 1990). At the same time, the escalating costs of waterway projects and shift of major funding responsibilities to local sponsors brought attention to a design process that compensates for uncertainty with conservative rules of thumb (Bertsche and Cook, 1980; National Research Council [NRC], 1983).

Port Development

Development of port infrastructure over the past 2 centuries has evolved through a balance of technological demands required by shipping, urban conditions affecting the port, and public interest in port modernization. Shipping technology has required increasingly deeper and wider channels and waterways (McCallum, 1987; NRC, 1981, 1985). Modern shipping terminals require larger spaces to operate and connectivity to greater and more-efficient intermodal land transport capacity. Larger terminals with higher volumes of cargo are in conflict with the vehicular congestion associated with ports encroached by or developed in urban areas.

Public support for port infrastructure modernization has softened. This decrease is due to competing demands for public investment funds for non-maritime-related purposes in the port area (for example, residential and commercial developments, recreation sites, marine habitat preservation, restoration) and to heavier emphasis on environmental aspects of proposed waterway projects than in former years, especially the impacts of dredging and the disposal of dredged materials (*Journal of Commerce*, 1991a; Kagan, 1990; Marine Board, 1985; NRC, 1981, 1985, 1987; Rosselli et al., 1990; U.S. Maritime Administration, 1990). Keeping pace with rapid changes in technology while keeping costs manageable and accommodating environmental interests of public policy and public interests groups has become more difficult.

The impact of the factors affecting modernization of the U.S. port infrastructure across the nation is selective. For example, some ports are experiencing extreme congestion on the land side due to differing urban conditions while other ports experience channel limitations on the marine side due to the timing of previous modernization. The overall result has been increasing demands on the local port and a lessening ability to solve port infrastructure problems.

Nevertheless, various modernization projects are in progress. They vary from a $5 billion port development proposed for Los Angeles-Long Beach Harbor to a wider and deeper channel in Miami; from a new container terminal in Tacoma to better rail access in New York. In any one year, over $500 million in port-funded capital improvements for port facilities and waterways is typically under way (*Journal of Commerce*, 1991b)

Water Resources Policy

Although funding development of port facilities is the responsibility of civil authorities and private enterprise, the federal government has historically led development of the port and waterway system (Heine, 1980; National Research Council, 1983, 1985). The major costs of construction,

WATERWAY TERMS AS USED IN THIS REPORT

BASIN A comparatively large space in a dock, waterway, or canal system, which is configured to permit the turning or other maneuvering of vessels for entering or departing a dock or berth.

BERTH A place where a vessel is moored at a wharf or lies at anchor.

CANAL An excavated, dredged, or constructed watercourse, usually artificial, designed for navigation. Side borders usually extend above the water surface.

CHANNEL Part of a watercourse used as a fairway for the passage of shipping. May be formed totally or in part through dredging.

DOCK The water space between adjacent piers or wharves in which vessels are berthed; an artificial basin or enclosure fitted with lock gates to retain a level of water undisturbed by entering or departing vessels (wet dock); any dock in or on which a vessel can be made to lie completely out of the water (dry dock).

FAIRWAY The main thoroughfare of shipping in a harbor or channel; although generally clear of obstructions, it may include a middle ground (that is, a shoal in a fairway having a channel on either side) suitably indicated by navigation marks (such as buoys).

HARBOR A fully or partially enclosed body of water offering safe anchorage or reasonable shelter to vessels against adverse environmental conditions. May be natural, artificial, or a combination of both.

PORT A place in which vessels load and discharge cargoes or passengers. Facilities in developed ports normally include berths, cargo handling and storage facilities, and land transportation connections. Normally a harbor city, town, or industrial complex.

WATERWAY A water area providing a means of transportation from one place to another, principally a water area providing a regular route for water traffic, such as a bay, channel, passage or canal, and adjacent basins and berthing areas. May be natural, artificial, or a combination of both.

WHARF A waterside structure, also referred to as a pier, at which a vessel may be berthed or at which cargo or passengers can be loaded or discharged.

SOURCES: Bowditch, 1981; McEwen and Lewis, 1953; Rogers, 1984; U.S. Navy Hydrographic Office, 1956.

VESSEL AND OPERATOR TERMS AS USED IN THIS REPORT

BARGE A heavy, non-self-propelled vessel designed for carrying or lightering cargo.

INTEGRATED TOW A flotilla of barges, tightly lashed to act as a unit. Common configuration found on shallow inland waterways.

PILOT The person piloting (directing and controlling the maneuvering of) the vessel. In actual vessel operations, the pilot could be a licensed independent pilot, master or qualified deck officer.

SHIP A self-propelled, decked vessel used in deep-water navigation.

TOW One or more barges or other vessels being pulled, towed alongside, or pushed ahead.

TUG, TUGBOAT, TOWBOAT A strongly built vessel specially designed to pull or push other vessels.

VESSEL A general term referring to all types of watercraft including ships, barges, tugs, yachts, and small boats.

SOURCES: McEwen and Lewis, 1953, Rogers, 1984.

---

operation, and maintenance of federal navigation projects, channels, and waterways used by marine transportation has been funded and built by the U.S. Army Corps of Engineers (USACE).

Until the 1970s, the federal government was the major source of funds for basic channel and waterway infrastructure, leaving actual port facility and land-side access up to local ports, other agencies, and private enterprise (especially for petroleum terminals). This redistribution of national resources directly benefitted local ports and their service areas, with indirect benefits accruing to the national interest in assuring the adequacy of the marine transportation system for regional and international commerce.

In the 1980s, federal policy changed. The shifting of more financial responsibility to local sponsors (for example, port authorities) began with the imposition of user fees. Substantially increased requirements for local sponsorship resulted from passage of the Water Resources Development Act of 1986. The act envisioned partnerships between the federal government and nonfederal local project sponsors in which local sponsors would have a significant role in planning, design, and funding. The federal government would continue to bear the major costs of basic waterways infrastructure, but local sponsors would be required to shoulder more of the cost for planning, construction, operations, and maintenance, and perhaps most importantly, responsibility for the disposal of dredged material. For example, local sponsors are required to share 50 percent of the costs of feasibility planning, provide the federal government with any needed real estate and property at 100 percent local sponsor expense, and contribute 50 percent of construction costs for the portion of project depths that exceed 45 feet.

These dramatic changes in federal policy have elevated the attention given by local sponsors to project costs. The changes have prompted interest in scaling down design dimensions to the minimum necessary for safe operations and minimizing the amount of dredging and the volume of dredged materials that must be disposed of. Traditional design methods using rules of thumb increase design dimensions to compensate for uncertainty. They assure adequate margins of safety but provide little comfort to designers charged with achieving maximum cost-effectiveness (Bertsche and Cook, 1980; NRC, 1983).

ROLE OF SIMULATION IN WATERWAY MODERNIZATION

The design of a waterway is as much an art as a science. Design must address many different qualitative as well as quantitative factors affecting its cost and operability. These factors include the engineering, operational, scientific, environmental, economic, and political aspects of a waterway project. Determining the swept paths of the vessels that will ply a waterway, for example, is an essential step in its design. These paths will reveal the relative risks of passage that must be addressed in waterway design. Characteristics such as channel depth, width, and geometry are selected in an attempt to optimize the balance between risk and cost inherent in the design.

A growing number of those involved in waterway design are applying high-technology systems to better determine the most cost-effective waterway configurations. One such technology, shiphandling simulation, has been used for operational training (for example, emergency procedures and maneuvering), analyzing marine casualties, evaluating vessel designs for maneuverability, evaluating bridge equipment, evaluating aids to navigation, and assessing the suitability of a particular vessel for a new port or transit situation. Shiphandling simulation techniques have also been used to select waterway configurations, usually as modifications to segments of an existing system, that accommodate economic, safety, and environmental interests. Additionally, available simulations have been used for multiple purposes of research, training, and waterway design (Ankudinov et al.,

1989; Burgers and Kok, 1988; Elzinga, 1982; Froese, 1988; Loman and van Maastrigt, 1988; McCallum, 1982; Paffett, 1981; Puglisi, 1987).

Initial shiphandling simulations involved remotely controlled scale models or scale models of sufficient size to accommodate human operators. In recent years, computer-based simulation has benefitted greatly from the advance of computer technology to help determine the swept paths of ships and integrated tows (oceangoing and shallow draft) under a variety of waterway configurations and operating conditions. [This report typically refers to ships, ports, and waterways for convenience of discussion. Integrated tow operating environments (such as found in river systems) are also assessed using computer-based simulations (Miller, 1979)]. Computer simulations can be performed using human pilots in a simulated ship bridge (that is, a functional mock-up) and mathematical models of ship behavior to predict the response of the vessel to commands from the pilot. Simulation is also performed in fast time using computer-based pilot models instead of human pilots.

Although computer simulations of both types have been used increasingly to aid in waterway design worldwide, there are many concerns about the practical application of the technology for this purpose. Widespread application has been hampered by questions of the validity and value of the results. This report assesses the validity of simulation as a design technique to better determine the feasibility, usefulness, and cost-effectiveness of computer simulations for the design process. It describes the waterway design process as it has traditionally been accomplished, the role of simulation in the design process, the components of a simulator, and the present state of practice. The application of simulators in several case studies is presented, and research needs are outlined.

2

Waterway Design Process

The design of a waterway is a highly complex and demanding exercise. The process involves an amalgam of economics, engineering, environmental and social aspects, political considerations, and historical precedence. Shiphandling simulation plays a role in a small but very important part of this process. This chapter describes the state of practice of waterway design to establish the context in which computer-based shiphandling simulations are applied. Also examined are typical design issues that appear to lend themselves to assessment through shiphandling simulation as well as design elements and data that are critical to successful simulations. The discussion provides a basis for understanding the advantages and limitations of shiphandling simulations and the potential for advances in the underlying technology.

The central role of the U.S. Army Corps of Engineers (USACE) in the waterway design process in the United States has produced a somewhat different institutional process (described in Appendix B) in comparison to the rest of the world. However, the engineering concepts used in waterway design are essentially the same.

A waterway design defines the form and dimensional boundaries required to meet functional objectives consistent with fundamental civil engineering practices and construction options. Construction includes excavation (dredging), manipulation of earth and rock, and the erection of heavy structures. As with other areas of civil engineering, the actual construction design is developed by the application of well-known principles of the physical sciences, such as hydraulics, geotechnics, and properties of materials. Most of the effort in waterway design has historically concentrated on these civil engineering aspects. Sedimentation has received particular attention, including the means to reduce or control it and its effects which might result from changing hydraulics within the waterway system relative to the tidal prism (National Research Council [NRC], 1981, 1987; USACE, 1977). However, the true challenge of waterway design is to balance the civil engineering requirements with those of form and function, including environmental considerations.

THE DESIGN CHALLENGE

The distinctive and unique thrust of waterway design is to quantify the factors that are used to determine the form and its dimensions for navigation. The process is difficult and involves complicated hydrodynamic reactions between the waterway and vessels. The difficulty is compounded by the fact that the vessels are independently controlled by human operators and sensitive to the pilots' reactions to varying operational demands. This fact, under any operating conditions, results in a certain lack of precision or certainty in determining vessel paths (Atkins and Bertsche, 1980; McAleer et al., 1965; Norrbin, 1989). Therefore, margins for safety need to be provided in the principal waterway dimensions. Estimation of appropriate margins is a key design function.

Principal design elements of form that are required to be determined and dimensions that must be developed for a given waterway are the following (Atkins and Bertsche, 1980; Dand, 1981; Marine Board, 1985; McAleer et al., 1965; McCartney, 1985; Norrbin, 1986; USACE, 1983):

- location
- orientation or alignment
- depth
- width
- radius of curvature of bends
- tangent distance between bends
- aids to navigation

These elements and their dimensions are primarily a function of the dimensions of the design vessel, its track, and its expected vertical and horizontal movement as it transits the waterway. Additional clearance dimensions to allow for uncertainty of position, operational safety, and hydrodynamic requirements are also required. Depth may also include a preinvestment factor to allow for sedimentation during intervals between intermittent maintenance dredging.

Vessel motions and path and clearance requirements can be estimated by calculation, physical tests, or semiempirical methods. The operating environment is so variable, and the calculations so complicated, that considerable judgment is usually required for design by traditional guidelines. In practice, it appears that most estimates have been made by applying semiempirical methods and judgment (Atkins and Bertsche, 1980; Dand, 1981; NRC, 1981; Norrbin, 1986).

Closely related to design is the operational analysis of a given waterway to appraise its capacity in terms of vessel size, traffic pattern, or density (Atkins and Bertsche, 1980). Operational analysis is applied to alternate design options, and the results are considered in optimizing design for safety and cost-effectiveness and when designing a navigational aids system. Operational analysis is also used by vessel operators (for example, shipping companies) to appraise the suitability of a waterway for a particular vessel and its loading limitations. A special case is forensic analysis where the conditions for an accident are deduced and re-created.

Much of the same technology is used for operational analysis as for design, but application techniques and methodologies may vary to reflect the somewhat different objectives. Acceptable tolerances for calculated results may also differ (Gress and French, 1980). For this study, operational analysis is considered a special case of design and is not explicitly discussed.

DESIGN ISSUES

The issues to be addressed in waterway design are both technical and institutional (Herbich, 1986; NRC, 1983,1985; Olson et al., 1986).

Technical Issues

Key technical issues include the following (McCartney, 1985):

• A design vessel or vessels must be selected with dimensions and characteristics around which the design is to be developed. The design vessel may be an existing vessel, a new vessel in planning or under construction, a conceptual ship of the future, or a composite of critical dimensions and properties of several vessels. Selection of the design vessel is a defining decision in the design process, regardless of design aids used (USACE, 1983).

• Dynamic behavioral characteristics must be determined for the design vessel (or vessels) as the vessel transits the waterway subject to various external forces and its own hydrodynamic and inertial properties. Related is the question of whether the vessel is to be maneuvered with or without the assistance of tugs (Armstrong, 1980; Brady, 1967; Crenshaw, 1975; Reid, 1975, 1986).

• The actions of a vessel's pilot must be determined relevant to dynamic behavioral characteristics (Armstrong, 1980; Crenshaw, 1975; Hooyer, 1983; Norrbin, 1989). Piloting skills resident in a local pool of pilots are not necessarily a critical factor in channel design studies. If a ship can be proved to be adequately handled by an experienced pilot, and thus the physics of the transit problem are not critical, then it follows that other pilots may be trained to operate the vessel safely, although better navigational aids might be required.

• Operating requirements must be noted, including the required speed of vessel in transit, density of vessel traffic in the waterway, traffic mix, special safety requirements, and degree of tolerance for risk of operating interruption (such as a grounding or collision).

• Assumptions for environmental conditions and limits must be assessed including oceanographic, hydrological (for example, tidal prism, currents, water levels), atmospheric, meteorological, and ecological factors, as well as time of day as it affects visibility.

• Costs must be determined for construction, maintenance, environmental and social impacts, and for vessel operations, together with their allocation and the assignment of benefits.

• Levels of risk that are acceptable must be determined.

Acceptable Levels of Risk

A special technical issue is risk. Tradeoffs made in design result in channel and waterway configurations that can be characterized as achieving an acceptable level of risk. There are no guidelines about what the acceptable risk level should be; thus the determination is highly subjective. During the assessment and based on its collective experience, the committee observed that port and public officials are reluctant to concede that some level of risk is an element in any port and waterway design. Reasons for this include concern over liability, project permitting, and interport competition. This general attitude has impeded the use of risk analysis with or without shiphandling simulation. Insight on risk can potentially be addressed by using simulation to identify maneuvering problems that may be associated with design alternatives or may be induced by certain physical conditions in the waterway environment. Moreover, the use of simulation has the potential to reduce the extra margins traditionally used to overcome uncertainty, thereby reducing construction and maintenance costs. Although this benefit may appeal to the project sponsor, it is the significant design refinement opportunities afforded by simulation that lead directly to the question of whether the fidelity of the technique justifies reliance on it over or in addition to traditional design practices to achieve adequate margins of safety.

Institutional Issues

Institutional issues are more difficult to define. Although they influence and are influenced by the technical issues, institutional issues are often the overriding and decisive factor in many waterway designs. They are usually associated with methods of finance, special environmental or social concerns, litigation, or legislation (Kagan, 1990; McCartney, 1985; NRC, 1987; Olson et al., 1986; Rosselli et al., 1990). Within the past several decades, competition for use of coastal areas has greatly increased. Competition exists between residential, industrial, recreational, and conservation uses. As a result, waterway development processes have come under much greater scrutiny by local interest groups, resulting in a lengthening of the already long approval process (Kagan, 1990; NRC, 1987).

In the United States, for example, the time interval between design study and construction for a federal waterway project is frequently more than 20 years (NRC, 1985), which means that the assumed technical issues will likely have changed greatly by the time construction is completed. Typically, the original design vessel becomes obsolete (and may no longer be in service), shipping practices change, new supportive technology is developed, and cost relationships are altered. There is no reliable methodology for projecting future vessel design or operational trends when planning waterways or for adequately accommodating changes that occur. Thus, original technical issues can be quickly overtaken by events.

Although an extreme, this situation in the United States is merely an exaggeration of global historical trends. Technological development in ships and in shipping operations have repeatedly stretched the technical limits and dimensional margins of waterways and harbors (McAleer et al., 1965; McCallum, 1987; Permanent International Association of Navigation Congresses [PIANC], 1980). The trend has been magnified during the past century and a half in response to the industrial revolution and expansive trends in world economic activity.

The dilemma for waterway designers is to balance the costs and benefits. Shipowners are the direct beneficiaries of increased efficiencies gained from larger vessels. Others may receive direct or indirect benefits but may also bear the costs of providing facilities (NRC, 1985), which affects the resources available for a project.

Institutional pressures and counter pressures (NRC, 1985) affect the designers' ability to implement a design of maximum utility and overall economic benefit. Ports and harbors are one element and represent a small share of the overall investment in the worldwide seaborne transportation system. However, construction, operation and maintenance cost are usually a major issue for local and national authorities responsible for funding. This fact, coupled with environmental and social concerns that have potentially significant cost implications, means that waterway development will inevitably remain under pressure for provision of minimal facilities or deferment (Kagan, 1990; NRC, 1985, 1987).

As a consequence, economic pressure to use ships larger than design ships into existing waterways will continue (Jensen and Kieslich, 1986). Because the United States has few natural deep-water harbors, waterway designers have had to continually reappraise vessel size and operating limits for existing waterways and have developed minimal incremental improvements for extending those limits, usually for economic purposes (Atkins and Bertsche, 1980). However, there has been no impetus for use of simulation for project-specific design reevaluation or safety appraisal. Assessing the effectiveness of waterway design once a project is constructed, the adequacy of the design for use by vessels exceeding design vessel characteristics, and the accuracy of simulation predictions are not elements of current practice. Furthermore, no one, including the Army Corps of Engineers, Coast Guard, or project sponsors, appears to be overseeing operations to assure that vessels using a waterway remain within the vessel operating characteristics for which the waterway was designed. In fact, there is an economic incentive for shippers and port authorities to exceed the design parameters of the waterway in order to accommodate the latest generation ships, thereby maintaining a competitive advantage relative to other ports and maximizing the amount of cargo that can be accommodated.

DESIGN PROCEDURES

The classic full-effort design procedure consists of the following steps (Dand, 1981; McAleer et al., 1965; Norrbin, 1986; Sjoberg, 1984):

- establishing various trial design alternatives to meet both civil engineering and navigation requirements;
- comparing their estimated capital and maintenance costs, benefits, and other factors; and
- selecting a best alternative.

Further incremental improvements to the selected alternative are usually considered and made by an iterative process until the design team is satisfied. By weighing tradeoffs in the cost-benefit analysis, the process approaches optimization (Burgers and Loman, 1985; Olson et al., 1986). True optimization of waterway design is seldom achieved or even attempted because of the lack of data, particularly data that are reliable and accurate and relating to accidents.

DESIGN PARTICIPANTS

CONGRESS Legislates project authorizations and appropriations for federal share of project funding.

U.S. ARMY CORPS OF ENGINEERS (USACE) Plans, constructs, and maintains federal projects in navigable waterways. Conducts technical research in waterway design and construction techniques.

U.S. COAST GUARD (USCG) Plans, constructs, maintains, and operates federal aids to navigation in navigable waterways; administers federal regulations pertaining to marine safety, security, and marine environmental protection.

U.S. MARITIME ADMINISTRATION (MARAD) Shiphandling simulation research and development; steering and maneuvering properties of ships. Provides advisory support in the design process.

PROJECT SPONSORS The local or regional organizations or authorities who contribute nonfederal funding to a specific waterway project. May include state, port, and local authorities.

LOCAL INTERESTS Segments of the local community with interests in waterway construction, operation, and maintenance who may act as petitioners or advisers for waterway projects. May include state and port authorities, terminal operators, shipping companies, and pilot associations.

PUBLIC INTEREST GROUPS Organized representatives from the public sector who have a direct or indirect interest in waterway projects. Interests include social, political, and environmental issues.

DESIGN ENGINEERS Technical design consultants to the USACE, sponsors, and other interested parties on a contractual basis. Principally involved in providing full technical support for waterway projects outside of the United States because most countries do not have the equivalent of the USACE and many foreign ports—especially in developing countries—are owned by private companies.

For public waterway projects in the United States, there is some attempt to follow this classic procedure, but with significant variations. The design process prescribed by the USACE has six phases:

- reconnaissance
- feasibility

SOME TECHNICAL TERMS USED IN THIS REPORT

DRIFT The sideways motion of a vessel from its track as it makes its transit.

DRIFT ANGLE The angular between a vessel's heading and its track.

SWEPT PATH A trace of the paths of the extremities of the vessel plan form as it makes its track while it transits the waterway. Account is taken of drift, drift angle, and yaw.

SWEPT PATH ENVELOPE The outer boundaries of the swept paths with the most extreme deviations from target track that encompass all of the swept paths of the vessels that transited the waterway.

TRACK A trace of the path of a vessel as it makes its transit of a waterway.

TRANSIT A passage of vessel from point to point in a waterway.

YAW The angular rotation of a vessel's longitudinal axis from the desired line of track.

- preconstruction engineering and design
- real estate acquisition
- construction
- operation and maintenance

The first two phases listed are theoretically where the waterway form and dimensions are determined (Olson et al., 1986). In practice, form and dimensions are fixed in the construction phase because of actual or perceived inadequate consideration of the interests of participants in earlier phases. In some cases, considerable delays in project approvals and implementation have occurred, resulting in a range of both constructive and detrimental effects (Kagan, 1990; NRC, 1985). A brief description of USACE design process mechanics is provided in Appendix B.

Developing a reasoned and sound technical design (which accommodates engineering, operational, safety, and environmental factors) as early in the process as possible establishes a solid basis for subsequent refinements. An issue is whether existing design tools are adequate to the challenge.

DESIGN TOOLS AND TECHNIQUES

The tools available to the waterway designer for technical solutions have improved markedly in recent years (Gress and French, 1980; McCartney, 1985; Norrbin, 1986; Olson et al., 1986). Before the ready availability of computers, designers were limited to carrying forward previous experience by judgment alone—with the aid of experiments with physical scale models—or by laborious mathematical calculations. Graphical methods with paper plots of position were often used to help visualize the pilot's task, the interaction of the forces on the vessels, and the vessel's resulting path. For practical reasons—usually monetary and time constraints—the number of model tests that were run and the variations that could be tested were usually limited. Similarly, the complexity of the mathematical solutions and applied formulae limited their number and required major simplifications to be usable. The calculations typically were used to check and verify previous assumptions rather than as a primary determinant.

The capacity and speed of the modern computer has changed the designer's task dramatically. Mathematical solutions are now practical from the initial stages of design. The relative ease of changing input conditions has broadened the feasible alternatives to be considered (Burgers and Loman, 1985; Gress and French, 1980).

Even with the modern tools available, the waterway designer must carefully input parameters and interpret results. To assist the designer, various groups, including USACE, PIANC, and International Association of Ports and Harbors (IAPH), have developed guidelines for design dimensions (PIANC, 1980, 1985; USACE, 1983). Although these guidelines are often helpful for visualizing a new waterway for initial studies, they are too general to assure an optimum design for a given condition. There are many examples of workable waterways that do not meet the guidelines by wide margins (Jensen and Kieslich, 1986; NRC, 1985). No substitute has been developed to replace intelligent and skillful analysis by a qualified, experienced waterway design engineer (Dand, 1981; Norrbin, 1986; Sjoberg, 1984).

Not all of the elements of a waterway are equally amenable to analysis by modern tools and technology. Basic data gaps and incomplete theories still exist. Although the technical press reports some study of the subject in recent years, considerable approximation and applied judgment are required for some elements and conditions. It is beyond the scope of this study to examine design factors in detail. However, several elements are important in considering the appropriateness of design tools and techniques, of which computer-based shiphandling simulation is one option. The elements are depth, width, aids to navigation, environmental data and civil engineering, and design vessel.

Depth

Depth is the key waterway dimension. It usually results in the greatest cost impact, establishes the character of the port and its traffic, establishes the initial and maintenance dredging requirements, and affects the horizontal controllability and resultant swept paths of the vessels that use the waterway. In the United States, project depth is determined by some technical analysis, but primarily by political and administrative means. Technical requirements for depth include allowances for (NRC, 1983; USACE, 1983)

- the vessel's expected draft;
- the vessel's vertical motion from squat or sinkage as it moves through the water and from pitch, roll, and heave caused by waves and other external forces;
- an under-keel clearance for hydrodynamic reasons; and
- an extra clearance to account for errors in measuring channel depth and vessel draft and for dredging tolerances. For new projects, an extra depth allowance may be included to allow for sedimentation to occur between intermittent maintenance dredging.

The primary technical tool for estimating depth requirements is designer judgment. Calculation of depth requirements involves the determination of critical sea and meteorological conditions, vessel operations, and other factors that affect the vertical motions and chance dimension errors. Because it is unlikely that maximum conditions for all factors will occur simultaneously, some designers have attempted to determine depth requirements by probabilistic forecasts. For example, studies for the Panama Canal Company in 1975 involved a special probabilistic approach related to pilot variance in compressed-time simulations (Norrbin et al., 1978). However, in practice, probabilistic forecasting has had mixed acceptance by designers. Even where practiced, considerable human judgment is still required for both input and evaluation.

Guidelines help, but in actual practice in many waterways, ship drafts consistently exceed those indicated as allowable by guidelines published by USACE, PIANC, and IAPH. Ships are routinely brought into ports with drafts that exceed project depths by taking advantage of daily tides and river stages (MacElrevey, 1988; NRC, 1983, 1985; Plummer, 1966). In practice, the only consistent, albeit informal, control over maximum draft on port entry or departure seems to be exercised by local pilots who make expert judgment calls on under-keel clearances that will permit safe movement of each vessel. Although the published guidelines offer a reasonable if imprecise gauge for safe under-keel clearances, economic criteria are applied by shipping interests.

Width

The required design width includes one or more vessel maneuvering lanes plus allowances for side clearances from the design vessel to the edge of the channel, other vessels, banks, structures, or natural features of the waterway. Width of the maneuvering lane is determined by the horizontal dimensions of the design vessel, its varying orientation in the waterway, and its deviations or drift from the desired track (Marine Board, 1985; USACE, 1983). A trace of the design vessel's extremities outlines and defines its swept path. The maneuvering lane is intended to provide an envelope of all the expected swept paths of the vessels that will transit the waterway under the various assumed design conditions. It is desirable that the lane's alignment is as close to a straight line as possible. Deviations in path alignment to avoid obstructions, take advantage of natural features, reduce dredging and sedimentation, or improve vessel operations are made with allowance for the design vessel's turning ability.

The side clearance dimension from an obstruction or bank provides a minimum path for the return flow of water displaced by a vessel as it moves along the edge of the maneuvering lane. It also provides a safety allowance for potential errors in the vessel's position. Deviations from desired orientation and vessel track are caused by a vessel's inherent stability or instability, the effects of external forces from wind, wave, current, and hydrodynamic reactions, and the applied control efforts by the pilot.

The degree of vessel control applied by the pilot is a major variable assumed by the designer. Unlike vertical motions, a vessel's horizontal motions and deviations can be anticipated and compensated for by pilot action. The effectiveness of this action is dependent on the pilot's level of skill, perception, and reactions, and following execution, on the inherent controllability and responsiveness of the vessel. Determining the degree of vessel control is a difficult challenge for the designer.

As with depth, actual practice has indicated that widths of much narrower dimensions than those recommended by traditional guidelines are both feasible and practicable. Some waterways such as the Houston Ship Channel fall into this category and are operated successfully (Jensen and Kieslich, 1986), although not without risk (Gates, 1989). Although technological gaps in the science still exist, there has been considerably more work done regarding width in recent years than there has been on the vertical phenomena. Calculation is feasible with a reasonable level of confidence.

Special cases, such as basins where low speed maneuvers are planned, bends and turns in channels, and passages through bridges, require special study. However, the design tools are generally the same and are available.

Weaknesses in the technological base include a lack of definitive data on maneuvering of vessels with very small under-keel clearances, especially in confined waterways. Also, quantitative guidance on the effects of different bottom material and contour forms and the effects of pitch, roll, and heave on track keeping in shallow water are scanty. Designer judgment and the transfer of prior experience are the principal tools to account for these conditions at present.

Navigational Aids

Aids to navigation systems are an important but frequently overlooked element of waterway design (Atkins and Bertsche, 1980). By integrating aids to navigation into the waterway design, the effectiveness and possible precision of vessel position fixing is improved and the designer can allow for tighter margins in waterway dimensions if they are validated by some means.

Available navigational aids range from traditional aids to navigation, such as buoys and ranges, to electronic position fixing devices, such as loran and differential GPS (global positioning system). All aids require human perception and reactions for maneuvering the vessel. Quantifying and evaluating behavioral modifications associated with use of aids to navigation is a particularly difficult challenge to the waterway designer.

Normal design procedure is to solicit the opinions and judgment of experienced mariners as a guide. Although this method frequently is satisfactory, it does not fully evaluate navigational systems in the context of a new or modified design. Shiphandling simulation has been applied and demonstrated to be of value for assessing aids to navigation (Atkins and Bertsche, 1980).

The waterway designer must carefully allow accuracy tolerances for behavioral modifications relevant to maneuvering strategies that may result from the type and placement of aids to navigation. Because unbroken delineation of channel boundaries and traffic lanes is typically not feasible in a waterway or fairway, the relationship of the vessel to its intended track is determined either by electronic or visual fixes (with some lag behind actual positions due to human and electronic processing time) or by expert estimations based on all information available. The pilot's strategy is therefore based on the perception of position and the onward track. Any lack of precision widens the track requirements.

Environmental Data and Civil Engineering

Navigational and civil engineering (including construction) aspects of waterway design require considerable data relating to the environment, both above and below the surface. Ideally, the data would be drawn from analysis of meteorological and hydrographic measurement records coupled with up-to-date physical surveys. Such records are not always available with the detail required for a specific project site.

Physical scale and mathematical hydraulic models have sometimes been used to interpolate general data regarding site-specific estimates of currents, sedimentation, and wave patterns. Similarly, mathematical models have been used with general synoptic charts for estimating meteorological conditions (Seymour and Vadus, 1986; USACE, 1977). Random or selective field measurements are usually advisable for verifying such estimates. As with all other aspects of waterway design, the engineering skill of the designer, together with clear and complete analytical reasoning, are prerequisites for success.

Design Vessel

Selection of the design vessel, or vessels, is one of the most critical decisions in waterway design (Dand, 1981; McAleer et al., 1965; USACE, 1983). Vessel dimensions and maneuvering characteristics are key to the required waterway geometry and dimensions, no matter what design method is used. The design vessel might be an actual vessel based on proposed operations or a hypothetical vessel. In accepted practice, the design vessel is selected to represent a combination of the largest ship with the least controllability that will require the greatest depth and largest width of the waterway, considering both swept path and clearances. It may not necessarily represent either the largest specific ship or the least controllable ship, although both are normally considered before a selection is made.

Ideally, vessel size and characteristics are based on forecasts of operations, considering world trends in shipping, and on forecasts of trade and traffic for the port. In actual practice, vessels used in most waterways differ substantially from what the designer had forecast 20 or more years earlier. In the committee's view, major reasons for this discrepancy include:

• dramatic changes in the form and composition of the national and worldwide merchant fleets made available through modern technologies;

• the time scale of the waterway development process, which is longer than the working life of a typical ship; and

• the absence of a waterways management regimen that restricts vessel access only to vessels that do not exceed design vessel characteristics (which could have the potential side effect of impeding development of maritime technologies).

Because of the inexact forecasts of future actual ships or vessels and the wide degree of variation in handling characteristics even of similar ships due to such factors as loading, the exact modeling of a particular design vessel is excessive. A reasonable approximation is sufficient. The objective is to model a representative vessel with typical behavior under the control of typical pilots under the conditions being studied.

Dimensions and other particulars for existing and new designs can be gleaned from naval architecture journals and publications from the classification societies. Estimates of the far future can be based on interviews with shipping interests and on deductive reasoning. In all cases, verification of handling characteristics by experienced mariners is of great assistance.

SUMMARY

Waterway design, whether for new construction, improvement of an existing waterway, or appraisal of the capacity of a waterway, involves estimating the navigation requirements of an assumed vessel or vessels, coupled with estimates of the civil engineering factors. Present technology allows calculation and mathematical modeling of the factors that affect waterway width and form in the horizontal plane, but considerable judgment still needs to be applied. Depth and other elements, including the need for aids to navigation, are still estimated and based primarily on human judgment. Human reactions by vessel pilots are an important ingredient, and their assessment and accommodation present a particularly difficult challenge to the designer.

Optimization of design, wherein all elements are appraised in terms of the others and alternate solutions are compared for maximum cost effectiveness, is not usually practical because of insufficient data and imperfect technology. Optimized designs in the United States are difficult to achieve because of institutional factors, such as increased emphasis on social and environmental objectives in design and the long lead times before implementation of a project after planning. Design tools or techniques are needed that can give reasonably correct technical solutions quickly and early in the process to provide a more scientific and technical basis for accommodating competing objectives that affect the waterway development process.

3

Use of Simulation in Waterway Design

Waterway design, which is reviewed in Chapter 2, is a complicated process. Some elements of the design process present opportunities for the application of computer-aided design techniques, including shiphandling simulation.

The use of shiphandling simulators to support the training of merchant mariners is generally well-known, and a number of ship simulators exist worldwide for training vessel operators and engineers. The emphasis of these simulators is more on reproducing the "feel" and behavior of the vessel rather than on predicting a vessel's trajectory with the accuracy needed for waterway design. Some simulators provide sufficient accuracy to accommodate both objectives. This chapter introduces the practice of using simulators to generate data that can replace or supplement "experience-based data" and rules of thumb, which have formed the basis for waterway design in the past. Simulation estimates the trajectory of design vessels that will use, or are projected to use, the waterway during its design life. Carefully designed simulator runs are used to gather the data that are then analyzed to draw conclusions about optimum or required minimum waterway dimensions and orientation, as well as ship operating procedures.

This chapter identifies the basic features of shiphandling simulators and simulation, the questions that simulation attempts to answer, and the basic assumptions that are made in simulation studies.

RELATIONSHIP OF PILOTING TO SIMULATION

A simplified block diagram of the full-scale piloting system is shown in Figure 3-1. The central component is the closed-loop feedback system consisting of the pilot, the display being used for navigation, and the response of the ship (those elements within the dotted box). The display represents the physical depiction of the present environment that affects piloting. The display can vary from a 360° visual view of the surrounding area on a clear, sunny day to just a radar image of the surroundings. The pilot interprets the situation and reacts by, for example, changing the rudder angle or increasing or decreasing thrust. Any changes in the heading and speed of the ship are discernable in the display. The behavior of this closed-loop feedback system is referred to as the behavior of a piloted ship.

Two other principal components in Figure 3-1 are the external environmental forces and the external visual environment. These blocks represent all of the external influences on the ship and on pilot behavior that are unique to the waterway, including channel topography, atmospheric visibility, tide, waves, currents, and wind, as well as the geographic features, such as aids to navigation, buildings, and bridges, that constitute the waterway

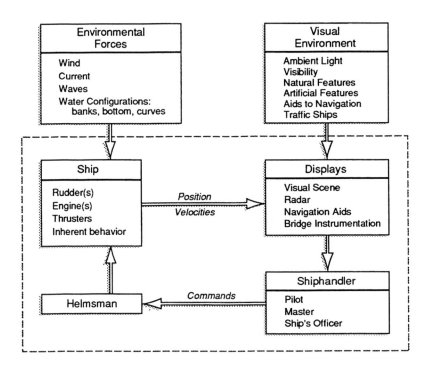

FIGURE 3-1   Block diagram of piloting.

environment. Some of these aspects are not fixed and can vary according to the moment (such as other ships in the waterway), the time of day (such as tide), from day to day (such as visibility, wind, and waves), or from season to season (such as flow in the waterway). The impact of many of these waterway features are generally known to the pilot only implicitly, that is, from the ship's reaction to them. Pilot ability to anticipate their effects is directly related to a pilot's familiarity with the vessel's operating characteristics and with the waterway.

The purpose of a simulator run for waterway design is to predict the track of a ship piloted by a mariner who is experienced in piloting in the existing waterway. Accordingly, a shiphandling simulator models the components of the full-scale piloting problem discussed above. The fundamental difference is that a simulator replaces the inherent behavior of the ship with an approximation of the behavior of a full-scale ship. This model of inherent ship behavior is a computer-based, mathematical model of the ship's dynamics. When possible, track plots of the simulated ship and the full-scale ship for the same maneuver are compared. The results are simulated trajectories of ship passages through a prospective waterway configuration in the same manner that the ship would be piloted under a variety of operational and environmental conditions if the configuration actually existed. Simulation can be accomplished using human pilots (real-time simulation) or using a computer-based pilot simulation (fast-time simulation). The mathematical pilot model used in fast-time simulation is often referred to as an autopilot, a term that can also refer to automatic equipment used to steer a ship on programmed courses or tracks.

Shiphandling simulators also include the other components shown in the simplified block diagram (Figure 3-1). Because the behavior of the ship is now represented by a computer model, the waterway must also be represented in a compatible (that is, numerical) fashion. Most shiphandling simulators include more extensive means of recording the results of pilotage than exist on an actual ship because the tracks and other information generated during the simulation are to be used in the waterway evaluation and design process.

Although modeling of ship behavior usually consists of a computer-based, mathematical model of the ship's dynamics, simulation facilities using physical scale models of vessels and waterways are also in operation. They have been used to aid in waterway design and results have been considered beneficial, particularly for addressing hydrodynamic factors. Physical model systems are generally constrained by physical and operating characteristics of vessel models on hand (or specially constructed for the simulation), waterway configurations that can be modeled at the facility, and if an outdoor facility, lack of control over external forces such as wind. Furthermore, the reaction times on reduced scale physical models are much faster than on the full-scale prototype. With regard to time scales, the pilot would in theory respond as if in real life but at an accelerated pace. However, artificial behavior could be induced through the ability to see quickly the results of maneuvering commands. The actual effects of all these differences on the faithful reproduction of ship maneuvering behavior by the pilot and resulting simulated trajectories are not known. As a result, most simulators developed in the last decade have been computer-based, which permits mathematical alternation of vessels and waterway configurations. These simulators are the subject of this report.

RELATIONSHIP BETWEEN SIMULATORS AND
THE DESIGN PROCESS

The many factors involved in designing a waterway, including civil engineering aspects, navigational aspects, and sociopolitical aspects, are discussed in Chapter 2. Indeed, a considerable history exists of designing waterways by design codes rather than by a detailed analysis. However, design codes are usually quite generous in their dimensions and undergo considerable refinement (and thus cost reduction) if credible analyses of the alternatives can be performed.

Fundamental to assessing how simulation can contribute to the design process is understanding the type of information that simulation attempts to provide. Fast-time simulation (also referred to as compressed-time simulation) provides the designer with many swept paths for the design vessels under a wide variety of conditions within the waterway (tide, current, wind, speed limits, and so on) in a compressed time frame. This information corresponds to, and replaces some of, the graphical constructions used in the simplified approach to channel design. Fast-time simulation can also be helpful in determining maneuvering lane width and overall waterway geometry early in the design process. Fast-time simulations are sometimes used to screen various design configurations for those that will be assessed through the more time-intensive real-time simulations. Real-time simulation (also referred to as full-mission simulation) uses qualified pilots to maneuver the simulated vessel through the modeled waterway using a true-to-life time scale. These simulations can be used in calibrating the pilot model for fast-time simulations, answering questions concerning navigational aids, and assessing piloting under difficult situations (complex bathymetry or environmental conditions, passing bridges or other marine traffic, and so on) where human decisions are critical.

The thrust of simulation in waterway design is to assess the risk to life, property, and environment of passage either for a new waterway or for an existing waterway (perhaps with new ships) without incurring either those risks or the costs of obtaining this information from real-life experience.

For convenience of discussion, assume that it is possible to make a simulator predict exactly the path of a given vessel in a given waterway with a given pilot. A program of simulated passages designed to provide a detailed assessment of risk of passage for the given waterway could, for example, involve simulating the voyage of every conceivable ship that would ply the waterway during its lifetime, under every conceivable state of the environment and traffic, and under the pilotage of all manner of pilots. Even if one could afford the cost of mounting such a program, the time required would be comparable to the lifetime of the waterway. Although this time would be significantly reduced if the real-time runs were mostly replaced by fast-time simulation, such a program would still be impractical.

The committee believes that applying simulation in waterway design relies on the following inherent working assumption:

> *A limited number of simulations using a less-than-perfect simulator, a few select (design) ship types, a few select environmental conditions over extreme ranges characteristic of the local area, and a few pilots with representative local expertise and shiphandling proficiency are sufficient to obtain a useful appraisal of waterway design.*

Evaluations of such simulations rely heavily on professional judgments and experience to identify or clarify design deficiencies, detect unforeseen problems, and determine areas where refinements would optimize the design to reduce costs without compromising safety. It may be possible to relax this assumption through the combination of real-time and fast-time simulations.

The validity of this assumption is critical to the efficacy of simulations as a design tool. Similar engineering assumptions are made in other fields with satisfactory results. For example, the design of an offshore platform requires the estimation of the worst loads that will be exerted on it during its lifetime (for example, loads experienced in a storm with a return period of 100 years). Statistical methods have been developed to estimate these loads from a limited environmental history and from limited model test results or analytical computations.

Conceptually, the undertaking of a limited program of simulations to appraise designs falls within accepted engineering practices. If a limited program is used, the relative accuracy or detail of each of the four elements (simulator, type of ships, waterway environment including vessel traffic, and pilots) must in some sense be balanced. The cost of a simulation program increases almost linearly with the scope of the program after the simulation model is set up. The design of a simulation program is therefore generally focused on determining the minimum scope of the simulation program necessary to make a meaningful risk assessment for a given design (or set of design alternatives). To accomplish this goal, the program is usually biased toward combinations of elements that will strain a waterway the most. The assumption is that if the waterway is satisfactory for these combinations, it certainly will be for other combinations that do not strain the waterway as much. Interpretation of the results must reflect the biases inherent in these choices. For instance, for many waterways the designer can anticipate traffic composed of a wide variety of ship types, some of which are not yet in existence. Some of the anticipated traffic may include small, maneuverable vessels that will ply the waterway with ease no matter what the waterway design; other traffic consisting of very large ships with limited maneuverability in restricted waters may strain the waterway depth limits, the maneuvering lane widths, or both. These latter ship types and cargoes carried often have potential for significant consequences (typically channel blockage or pollution) should accidents involving them occur.

Typically, research is directed toward the application of a specific ship to an existing waterway or a waterway to be constructed. In other cases, when many different ship types are involved, the selection of the ship or ships to be used in simulation is subjective, relying heavily on experience. Ideally, the selection is based on the input of pilots who are familiar with the area of the proposed waterway and who are qualified to pilot the types of ships to be simulated. If local pilots do not have experience with the simulated vessel, pilots from other areas with the necessary ship maneuvering expertise could be included in the study. Ship selection must also involve some description of the loading conditions of the ships, because the behavior of a fully loaded ship with small under-keel clearance will likely be very different from that of the same ship, lightly loaded, with a large under-keel clearance and more subject to wind loadings.

Designers may also anticipate and design for increased risk of an accident during severe environmental conditions (for example, storms, high currents), which could severely strain the skills of even the most experienced pilot. Like the selection of ships for simulation, selection of these additional factors ultimately is made subjectively. In the past, many questions and some controversy have arisen about what can be reasonably assumed for pilot control and skill in the selection of weather conditions, aids to navigation, and dimensions of waterways. This uncertainty is especially true when estimated ship trajectories are developed by simplified analytical schemes that do not put qualified pilots in the simulation process. The same controversy also applies to trajectories estimated from fast-time simulations. By using experienced ship handlers in a real-time simulation and presenting them with an adequately realistic situation, the question of applied skill level of the pilot is addressed, if not fully answered.

Real-time simulation with human control is gaining acceptance throughout the world as a useful aid in harbor and waterway design. Some of the many applications to date are discussed in Appendix C. Although its acceptance has been slowly and steadily increasing, there is no consensus amongst designers concerning its usefulness, even though it has been used as a tool in early development stages of some waterway designs (Norrbin et al., 1978; Ottosson and van Berlekom, 1985; Puglisi, 1988; Simoen et al., 1980). Reasons for the apparent reluctance to use simulation for concept development include:

- cost and time requirements;
- the validity of the modeling; and
- interpretation of results.

In addition, some waterway designers may not be comfortable with changes from traditional techniques with which they are very familiar to a process that not only may not be familiar to them but also would expand participation beyond the traditional design community.

Because real-time simulation is human resource intensive, the capability for quickly modifying inputs to the mathematical model which describe the waterway and its environment is desirable to facilitate assessment of design alternatives. There sometimes is difficulty in achieving this objective depending on the waterway under examination. From an examination of several case histories where simulation was used (see Chapter 7; Appendix C), it appears that these objections are not exaggerated, although the difficulties did not prevent project sponsors from acquiring valuable technical and design data. In time, as users become more familiar with the tool and its use is refined, simulation may play a more important role in design, particularly much earlier in the process.

Special design problems for which real-time, human-controlled simulation appears particularly suitable are the following:

- determining a pilot's ability to assess the vessel's position in relation to horizontal dimension requirements, including the value and placement of navigation aids;
- evaluating traffic density limitations;
- optimizing side clearance dimensions for a vessel of a given size;
- maneuvering actions, including docking and undocking; and
- optimizing bend and turn dimensions for a vessel of a given size.

All of the above considerations are important in waterway design, and all are almost totally dependent on applied pilot skill. Heretofore these problems have been addressed mostly on the basis of opinion without a means of quantification other than full-scale testing.

Sometimes the resolution of design problems has been as much political as technical, necessitating extensive efforts to achieve a consensus between parties with conflicting views. The committee found that where used (see Chapter 7; Appendix C), simulation has been a unique way to test opinions on specific designs in a focused and clearly visual way. Furthermore, real-time simulation has in some cases helped to build consensus in the design process by providing a realistic presentation of problems that is understandable to all interested parties.

SUMMARY

Simulation is a technology used for predicting the track of a ship in a waterway either by using qualified pilots (real-time simulation) or a pilot model (fast-time simulation). Typically, simulation runs primarily reflect situations that will most stress the waterway and the number of these runs that can be made is limited. Nonetheless, useful technical information concerning the vessel track can be obtained, and consensus building among the conflicting parties in the waterway design can be achieved.

4

Shiphandling Simulators

Shiphandling simulators encompass a wide range of capabilities, facilities, and man-machine interfaces. They can be divided into two major classifications: real-time simulators, which have a human controller (referred to as a *man-in-the-loop*), and fast-time simulators, in which the human is replaced by a computer-based pilot model (often referred to as an autopilot). Because there is no human involved in fast-time simulation, the speed of the simulation is limited only by the speed of the host computer. With modern computers, these simulations can be performed at much greater speed than real time.

Simulation allows examination of proposed waterway designs before they are selected or implemented. The primary contribution of simulation is quantitative performance data characterizing the design and operational alternatives being considered. A number of methods furnish data that can be used in the design process: physical models, fast-time mathematical models, and man-in-the-loop simulation. The latter provides data on the entire navigational system, including the variability of the shiphandler, and thus it is referred to as *full-mission* simulation. This method provides subjective evaluations as well as quantitative assessments that can be used to guide the selection process and acceptance of a proposed design.

Both fast-time and real-time simulations are available offering various levels of sophistication. The phenomenal growth in computing power and its low-cost availability relative to the total cost of a simulator program has eliminated simulator hardware as an important limiting factor in applying simulation to waterway design. Instead, the cost of obtaining data for the mathematical models (called *identifying* the model), processing of these data, and developing the visual scene are emerging as the dominant costs in marine simulation.

COMPUTER-BASED MODEL FOR SHIP BEHAVIOR

A simulation model for ship behavior is a computer-resident mathematical model of the waterway and of the dynamic properties of the ship. The waterway model includes not only the bottom topography, but also the winds and the currents below the water surface. This model is usually a combined data base and interpolation scheme where model details can be determined for an arbitrary location in the waterway.

The core of the ship dynamic model is the set of equations of motion of a rigid body (the flexibility of the ship is inconsequential for these problems). The equations of motion are usually referred to a coordinate system fixed in the ship, and the result is called Euler's equations of motion. These equations are six-coupled, nonlinear, ordinary differential equations that relate the motions of the ship to arbitrary external forces acting on the ship.

The force system that acts on the ship is a result of hydrostatics, hydrodynamics, and aerodynamics. Hydrostatic forces can be computed by Archimedes' principle. Techniques for numerically modelling the exact flow of air or water around a ship hull do not exist, even in the open ocean where the topography of either land or ocean bottom is not an influence. In a waterway, these local topographical details are very important and strongly influence the hydrodynamic and aerodynamic forces acting on the ship. As a result, a combination of theory, experimental results, and heuristic approximations is used to determine mathematical expressions for the force system on a ship in a waterway.

In addition to the ship's dynamics, mathematical models are developed for several other dynamical systems. These include the main propulsion system; steering machinery; thruster machinery, if available; and assistance of tugboats.

The success of the computer model in reproducing a vessel's behavior depends on the ability to describe the waterway and its environment numerically, to predict the instantaneous force system on the ship, and to integrate the mathematical expressions or algorithms of the ship and other mechanical components that contribute to the vessel's trajectory. Each of these elements involves approximations, and in the end, each is reduced to a set of equations. A detailed discussion of this process is given in Chapter 5.

In recent years, digital computers have been used exclusively for these problems. Inputs to the simulator computer are the commands issued by the pilot, and output is the position and velocity of the ship and the visual presentation. Because the size (and cost) of modern digital computers of high capability is so low, the limitations on speed and memory capacity that were important considerations 20 years ago no longer exist. Therefore, the quality of the module for the vessel's behavior rests solely on the expressions, algorithms, and data bases that are programmed into the computer. In practice, these algorithms, hull and model scale data, and expressions are usually proprietary to the individual simulation facility, and as a result, comparisons of these aspects among facilities are limited.

Specific Components of Fast-Time Simulations

Fast-time simulation is closed loop, with a pilot model rather than an actual pilot in the loop. The pilot model, or autopilot, is computer software that simulates, to some level, the human performance of a shiphandler. It is not to be confused with the autopilot hardware found aboard ships that steers set courses or predefined turns. Pilot model software provides dynamic access to the necessary vessel motion and waterway data base parameters and algorithms to evaluate the vessel's track and to generate appropriate control commands for the vessel.

The typical autopilot for fast-time simulation is part of the software in the computer-based simulation and is defined in terms of the algorithms it uses to evaluate the vessel's track and to generate control commands. In typical fast-time simulation, the simulator operator supplies additional information in the form of a preferred track for a point on the ship (usually at midships centerline). The track may also include desired speeds for various parts of the transit that are used to trigger engine commands. This track therefore represents a predetermined strategy for negotiating the passage and is presumably geometrically feasible (that is, if the ship follows the path exactly, no portion of the ship should extend beyond the bounds of the waterway). The pilot model in this situation performs as a track follower, because this approach roughly approximates what happens when a human pilot starts a passage with a set strategy and adjusts the transit when deviations from the planned route occur. In practice, the programmed track for the pilot model is usually selected after consultation with experienced pilots about appropriate transit strategies.

Autopilot designs can vary in sophistication depending on their needs. At one end of the spectrum, a simple autopilot is used. This autopilot, using the exact location of the ship (as computed by the mathematical model) as datum, generates simple rudder and engine commands as specified by the associated transit strategy in an attempt to minimize any deviation between the current location and the prescribed track.

The simulation obtained by simple pilot models are useful because dynamically feasible swept paths can be defined. For example, results can include deviations from the desired path resulting from such factors as inertia of the vessel, hydrodynamic effects, and lags in the power plant and steering gear.

Attempts are often made to make the pilot model behave more like a real pilot, including one or more of the following refinements:

- making the position error zero until a certain detection threshold is met,
- not executing a command until it is a significant one (for instance, waiting until a rudder command of at least 5° or 10° can be given),
- providing for anticipated course changes,
- introducing delays in decision making, and
- introducing noise (random error) into the position information.

Each refinement generally leads to a different swept path, which is in theory reflective of the swept paths that would be experienced with human pilots whose performance varies due to many factors such as professional skills, experience, and stress (including fatigue, boredom, and other physical condition factors). A virtue of the pilot model is that its performance can be made consistent (that is, human variations are screened out) so that the actual effects of physical forces on the design vessel for various tracks can be assessed through sensitivity analysis. Data from each approach can also provide a comparative basis for accommodating operational factors in the design.

Although the measures for representing human behavior introduce some variability into the pilot model, they do not achieve any semblance of the full complexity of human pilot behavior that reflects many different styles and levels of effectiveness in shiphandling. To represent the underlying perceptual and cognitive processes involved in detecting and interpreting aids to navigation and vessel traffic, decisions about maneuvering actions, and other operational decisions, much more sophisticated pilot models than exist today would be needed. Only then could true transit strategies be programmed into the simulation without taking the vessel into some predefined track. Potentially, developments in piloting expert systems (that is, computer-aided, knowledge-based decision making including use of neural networks) could be incorporated into pilot models to reproduce some of the more cognitive aspects of piloting behavior (Grabowski, 1989).

Fast-time simulations are usually used for sensitivity analyses because they do have consistency. One typical use is determining the effects of current variations and tidal stages on maneuvering. Because many runs can be performed in a reasonably short time, many different hydraulic conditions can be used. Another use of fast-time simulators is to evaluate a select number of alternate waterway designs. In either case, detailed records of the commands and resulting trajectories must be kept for analysis. This record-keeping requirement applies to real-time simulations as well.

Specific Components of Real-time Simulators

Real-time simulation involves a number of large and often expensive physical components that are not used in fast-time simulation. These simulations must be run in real time because they involve the participation of the human pilot to interpret the progress of the transit and to issue commands. Paths of communication must be provided between the pilot and the computer, including a means of displaying the location of the ship to the pilot (visual display) and a means of communicating pilot commands to the computer (controls).

Visual presentation

Two different types of visual presentation of the vessel's situation in the waterway are common. One corresponds to a *bird's-eye* (plan) view, such as a radar screen; the other corresponds to a bridge-view display that resembles what the pilot might see looking out from the vessel's bridge. Of the two display systems, the bird's-eye view is by far the simpler one to develop and requires only modest computational capacity. The bird's-eye or situation display is often more detailed than the corresponding radar scene and may include an accurate depiction of the vessel, geographical landmarks, aids to navigation, and other waterway features. Simulation of the corresponding radar image can be effected by eliminating or reducing much of this detail. When coupled with information equivalent to what would normally be available on the vessel being simulated, this display can create a simulated operating environment corresponding to restricted visibility atmospheric conditions. Regardless of the display format, research has determined that if a simulation system provides more information to the pilot than available in real operating scenarios, the results of simulation may not be representative or useful (Norrbin, 1972). The results can also be biased if important information is missing.

Bridge-view displays are intended to be viewed and interpreted by the pilot as a representation of what would be seen during vessel transits when the atmosphere does not completely obscure the view of landmarks and aids to navigation. This display could simply be one monitor (corresponding to what might be seen out of one bridge window) or an array of screens presenting the pilot with a rendering of an actual 180° or a full 360° view. When using these displays, the following interdependent physical factors limit perceived realism:

- display size,
- physical field of view,
- viewing distance (from the eye), and

•   display quality (for example, color resolution, spatial resolution, brightness, and contrast ratio).

Bridge-view and bird's-eye displays are usually computer generated and are presented using either cathode-ray tube (CRT) monitors or large-screen, television-like projectors. When projection systems are used, the display size can be made fairly large with increased eye-to-screen distance for realism. However, the resolution does not improve, and the brightness decreases in inverse proportion to the area of the display. A compromise between brightness, screen size, resolution, and eye-to-screen distance can be made to present reasonable visual cues to the pilot.

The physical field of view refers to the angle subtended by the display as seen by the person performing as pilot during the simulation. Usually this angle is measured in the horizontal (azimuth) plane for ship maneuvering problems. A small television-size monitor can have a large field of view if it is placed close to the observer. However, viewing such displays can be uncomfortable, and they do not impart the feeling of a view from the vessel's bridge or pilothouse due to eye-to-screen distance. For realism, it is important that the field of view being represented in the simulation scene be approximately matched by the observer's field of view of the simulation display. Both the observer's and the simulation scene's fields of view from a single monitor or projector are inherently limited. Large display systems are often composed of three, five, or more display screens arrayed in roughly a circle around a mock-up of the bridge. Coordinating the projectors for multiple displays is not difficult with today's technology; it is possible to obtain displays with up to 360° of azimuthal field of view, although a somewhat smaller field of view, about 240°, is more common. Vertical fields of view vary depending on the simulator application, 20° to 24° being typical. Docking simulators generally require a larger vertical field than those applied to maneuvering or channel design work. Reasonable depth perception and reduction of parallax error for the simulation scene in relation to a simulator's bridge typically require a screen-to-eye distance greater than 10 feet. The closer this distance is to real-life conditions, the smaller the parallax error.

Display quality can be measured by a number of factors, including resolution, update rate, and texture. Spatial resolution refers to the fineness of detail that can be displayed. For computer-generated displays, the smallest unit of display is called a pixel. Individual computer displays are generally rated by the two dimensions of the array of pixels forming the display. However, what is more important for a simulator display is the visual angle a pixel subtends for a pilot located on the simulator's bridge relative to the angular visual acuity of the pilot's eye. Depending on the sophistication of the computer display generation, the appearance of any pixel can be chosen from a limited number of colors or greys, or from millions of different colors to reproduce natural shading and texture.

Update rate is the frequency with which new scene information is displayed. A slow update rate makes the scene "jump," whereas a fast update rate (greater than 15 hertz) yields a movie-like smoothness of motion. This rate depends critically on the computer supplying the graphical information and is often significantly slower than the refresh rate, the rate at which the screen is "repainted" by the electronics. Higher refresh rates eliminate screen "flicker" which contributes to viewer fatigue. The computer determines what is displayed by computing a two-dimensional perspective view of the scene as observed from the pilot's station. The computer derives this view from a three-dimensional description of the modeled environment stored in its memory. The speed of the process depends on the computer's ability to form the elemental shapes comprising its two-dimensional picture, to eliminate hidden lines or surfaces, and to determine the color of each pixel. Special, dedicated computer systems have been developed to perform this type of calculation with great efficiency. Because the update rate varies inversely with the number and type of objects arrayed in the three-dimensional space that will be visible or bounded within the scene, simple scenes can be updated at a faster rate than complicated scenes.

An elemental shape formed in the perspective view (e.g., a polygon representing a buoy) can be filled uniformly with the same color or filled with different colors which form a pattern reflecting its "texture" or shading (where, for example, the smooth texture and shading of a buoy may differ from that of the surrounding water surface). Many newer graphical computers have the capability to produce such texture, which can contribute significantly to the apparent realism of the display. A special use of texture is the "greying" of distant objects to enhance the observer's feeling of distance.

Controls

Another aspect of the physical setup of a real-time simulator is the realism of the controls and the navigation instruments used in the mocked-up ship's bridge. There is a wide range of mock-up realism in common use for simulators. In the most modest facilities, the only display may be a single CRT monitor for radar and visuals; the controls may be simply "radio" knobs that can be turned to give commands to the engines or rudders; and the navigational instrumentation readout may be simply a printer or portion of the CRT that shows the current readings. In the most elaborate facilities, the complete bridge of a ship is duplicated including all standard, commercial instrumentation and controls. The equipment, furnishings, and bridge windows are arranged to conform to traditional bridge layouts or layouts specific to the design vessel under study. Some facilities mount the bridge on a motion platform and include loudspeakers. The objective is to enhance the *fidelity* of the simulation by providing an approximation of the sound and feel associated with vessel response to environmental factors (for example, pitch and roll in a seaway) and maneuvering commands.

Fidelity

The word *fidelity* in this report refers particularly to the appearance and functionality of the simulator as experienced by the pilot. In the literature, the concept of fidelity often includes separate measures for various other components of simulation (for example, the mathematical model).

Ideally, the pilots are provided an environment that so closely resembles a ship's bridge (or pilothouse) that they are unable to detect that they are not aboard ship. In other words, the ideal is a bridge that looks, smells, feels, moves, and sounds like a real ship's bridge and has views through the windows and ports that are absolutely lifelike. Such an environment would be referred to as having "perfect" fidelity. These environments are, in fact, almost achieved for the training of aircraft pilots. The quality of the display and the realism of the mock-up contributing to fidelity are directly related to their costs, although the costs of the display hardware have dropped dramatically with advances in computer technology.

The actual environment presented to pilots in a simulator inevitably falls short of perfect fidelity, varying considerably from facility to facility. Most simulator facilities attempt to include appropriate displays that either mimic those on board an actual ship or at least evoke their presence. Some simulators incorporate the angular motions resulting from seaway and maneuvering by tilting the bridge and display systems. These motion systems can be extremely costly. Although one might naturally assume that higher fidelity is better, evidence is lacking that correlates the influence of fidelity to the results of the simulation. Consequently, no consensus exists among simulation practitioners regarding what levels of fidelity are required to achieve reliable simulation outcomes, or how the requirement might vary with the simulation study objectives.

There is considerable interest in the potential of technical representations such as electronic charts and real-time positioning displays to substantially augment and perhaps become more important than visual observations. A large number of performance and application issues are being researched, including the effectiveness of integrated displays for use in piloting waters (Astle and van Opstal, 1990; Clarke, 1990; DeLoach, 1990; Eaton et al., 1990; Grabowski, 1989; Graham, 1990; Kristiansen et al., 1989; Maconachie, 1990; Russell, 1987; Sandvik, 1990). However, there are many national and international policy issues requiring resolution. Thus, adoption of advanced systems cannot be reliably forecast.

The eventual use of integrated bridge displays and the subsequent changes these may have on piloting, will need to be reflected in future real-time simulations. Since it is less expensive to emulate a high technology bridge than to produce a high fidelity visual scene, it is likely that the cost of these changes would not become an issue. For waterway design, it would be prudent to equip real-time simulators with display systems of appropriate fidelity and bridge equipment that reflect the probable state of practice for merchant shipping. Improved performance by ships with advanced positioning and control systems could be expected to provide additional margins of safety relative to simulation results.

Man-in-the-loop

Pilots as shiphandlers represent the most complicated element in the behavior of ship maneuvering through a waterway (see Figure 3.1). Pilots must integrate diverse information acquired via the human senses on all aspects of own ship, environment, and other vessel traffic, as well as navigational conventions and other factors (Armstrong, 1980; Crenshaw, 1975; Plummer, 1966). Because pilots are integral to shiphandling in confined waterways, inclusion of pilots with knowledge of the design vessel, local conditions, and tug assistance is essential to the simulation process, if the simulation is to have complete credibility with prospective users of the waterway.

The pilot views the waterway scene from the bridge directly or, especially in the case of obscured vision, through electronic means including VHF communications, radar, and electronic aids to navigation. The pilot gains immediate information about the vessel from the readouts of the vessel control instruments. Based on the pilot's experience on the waterway and interpretation of the situation, rudder and engine commands are given. The pilot's skill involves the ability to determine the vessel's position and motion within the waterway based on observation, to predict change in the vessel's track resulting from the local environment, and to initiate required maneuvering commands in anticipation of the vessel's progress so that it will remain on the desired track.

Selecting appropriate pilots to participate in a simulation involves considerations of piloting skills, local waterway operating practices, and statistical sampling factors. Even when pilots for a simulation study are selected as a representative sample of the local pool of pilots, significant variability among pilots and their piloting performance is inevitable. Piloting different types and sizes of ships in waterways is a skill that takes many years to learn. Piloting skills vary with the nature of service (that is, coastal, bar, river, harbor, or docking pilotage, or a combination of pilotage services), experience, and personal capabilities. Also, a pilot's skill level may vary from day to day depending on human and external factors.

Piloting in a shiphandling simulation is somewhat different from real piloting. During the design stage of either a new waterway or an extensive modification of an existing waterway, no pilots have had experience with the design. The generic skills required to pilot a simulated ship through the waterway are, however, identical to those required if the waterway design is executed. Thus, it appears sufficient to use capable pilots for real-time simulation, with the understanding that these pilots will probably need to become familiar through simulation with the new or modified waterway. It is equally important that these pilots are familiar with the local operating environment and incorporate local knowledge into the simulation. Often, this can be assured by using pilots certified by the Coast Guard or appropriate state or local authority for the pilotage route.

If simulation runs are performed by pilots whose ability far exceeds the expected average for the waterway under examination, the results of the simulation may be overly optimistic. The converse may apply for simulations run with very capable pilots who lack the knowledge of local conditions. The fact that pilots know they are only performing a simulation and that the consequences of failure will not include vessel damage, lawsuits, or personal injury, may make their performance quite different from real-life pilotage.

Some simulator facilities always use a few select pilots in rotation. In these circumstances, peer pressure to excel and the simulator's sophistication may affect performance. This may be an advantage for simulation training but not for waterway design where duplication of real life performance is needed. Because there are no universally accepted measures of piloting skill or knowledge, assessment of these dimensions is subjective. It is difficult to evaluate whether or not pilots chosen to participate in a given simulation reflect the average capabilities of local pilots or how their performance may have been affected by simulation conditions or the pilots' sophistication with simulation techniques.

LEVEL OF SIMULATION

Each simulation facility conducting port and waterway design work uses different simulator components. These facilities are often compared by characterizing their components using a subjective measure called the *level of simulation*.

To indicate how this subjective measure might be arrived at, Figure 4-1 shows the various components that make up a real-time simulator and further subdivides these into the characteristics of the components. A profile

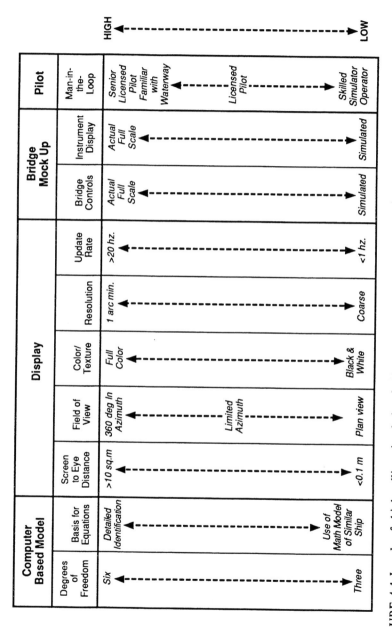

FIGURE 4-1 Levels of shiphandling simulation for evaluating real-time simulation facilities. Each column represents a scale for an individual characteristic of the simulation facility. The most desirable characteristics are at the top of each column, and characteristics of decreasing complexity (and generally decreasing cost) are listed down each column.

of any facility can be made by placing a horizontal mark in the figure at a height for each component that approximately reflects that facility. If the marks generally lie near the bottom of the columns, the facility is referred to as having a low level of simulation; if the characteristics generally lie near the top of the columns, the facility is referred to as offering a high level of simulation.

Practitioners of simulation generally accept that various levels of simulation are appropriate for different design situations. No definitive guidance is available to assist prospective users in determining the level of simulation needed for a particular problem. There is a dearth of quantitative information relative to selecting the level of simulation appropriate to a particular waterway design study. Nevertheless, if the level of simulation is not sufficient to capture an essential feature of the waterway, ship dynamics, or other key aspect of the real system, then the results of the simulation may be suspect. In practice, a higher level of simulation than what appears necessary is often used simply because the consequences of overlooking some subtle feature may have an important impact on vessel transit results. For example, the presence of a full bridge team to provide navigational support to the pilot would add to the face (that is, apparent) validity but would not necessarily add to the level of simulation. The actual contribution would depend on the capability of the bridge team to assess the operational situation and communicate this effectively. Thus, as with waterway design generally, the tendency in simulation is toward conservatism.

After the vessel (or vessels), environmental conditions, and appropriate simulator hardware are selected and installed, the simulation process occurs in three steps. In the preliminary phase, mathematical models of each simulation component are collected, and the various constants are identified (see Chapter 5). In the simulation phase, the model is exercised in either real time, fast time, or both. In the interpretation phase, the simulation results are assessed in terms of the risks posed by the channel design and potential alternative designs. However, because the simulation program is biased toward the most accident-prone situations, results must be carefully interpreted (see Chapter 6).

SUMMARY

The prospective user of shiphandling simulators for waterway design is confronted by several factors that complicate the decision to use a simulator. Given the range of technical considerations, careful examination of the capabilities, research methodologies, and results of available simulations is needed to assess simulator suitability for each individual waterway design project. Of equal importance is the selection of pilots for real-time simulations because they are critical to both the validity and credibility of results.

5

Mathematical Models

Many components of shiphandling simulators are substantial physical pieces of hardware. Some components can be evaluated easily from their appearance (the bridge and its equipment) or performance (the size, resolution, or update rate of the display). The mathematical model, which is embedded in the simulation computer and invisible to the user, is difficult to generate and even more difficult to validate. This section describes the state of practice of the development of the computer-based model for a shiphandling simulator. Validation of the model is presented in Chapter 6.

SELECTING AND IDENTIFYING THE SIMULATION MODEL

Before a simulation can be performed, it is necessary to develop quantitative computer-based models for the waterway, ship, and various components of the traffic. Each of these models consists of two kinds of information:

- a *framework* (or structure) for the data (which describe the generic component), and
- a set of *numerical constants* associated with the framework.

The framework is a widely applicable mathematical procedure or algorithm that embodies the relationships between the various factors involved. Numerical constants or coefficients quantify these relationships for the specific case under consideration. Selection of a particular framework for a specific component varies from facility to facility, is usually based on theoretical developments associated with that component, and is usually proprietary. Determination of the numerical constants associated with the framework is called *identification*. A discussion of both the selection of a framework and identification of its constants for each major component of the computer model is presented in the following sections.

WATERWAY BATHYMETRY

In order to determine the forces that act on a ship, it is necessary to determine the bathymetry (depths, contours) of the waterway in the neighborhood of any position the ship might assume during its passage. The framework typically consists of a data base that stores the waterway depth at specific locations and an interpolation scheme for these data that allows estimation of the water depth at an arbitrary point in the waterway. The structure of these data bases varies considerably. No clear advantage has been demonstrated for any particular scheme. Usually selection is a tradeoff between size of the database and ease of interpolation, which translates into a tradeoff between the storage capacity and computational speed of the computer used for the simulation. The presentation of the data in the original source is a strong influence on the selection of framework.

A typical framework for the data base is a grid on a chart of the waterway (either a rectangular grid or curvilinear grid fitted to the channel). Entries in this matrix correspond to water depth at each node of the grid. Data points must be specified with sufficient density to capture the underwater geometry of the waterway. Of the various choices, the rectangular grid (normally based on latitude and longitude) requires the largest number of data points but is simplest for interpolation. Data bases that use a waterway-fitted grid (for instance, one that uses the channel centerline as one coordinate) are much smaller but require more complex interpolation.

In any of the grid data base systems, different levels of interpolation can be used. Linear interpolation is the easiest and has the advantage of being most computationally robust. However, linear interpolation is also the least accurate because the interpolated values always lie within those data base values used as input to the interpolation. Higher order schemes, such as parabolic interpolation or cubic spline interpolation, require fewer data base points. However, if the data base points do not correspond to a smooth surface, anomalous interpolations can occur. Consequently, linear interpolation is most often used.

Some facilities use a different system altogether, one in which the numbers stored in the waterway data base correspond to polygonal contours of equal draft. Although this scheme results in an extremely compact representation of the bathymetry of the channel, it also results in the most computationally demanding interpolation scheme. The bathymetry of very complicated waterways can be described to the degrees of accuracy necessary with either the grid system or the contour system. Increases in accuracy require corresponding increases in the amount of stored data in the data base regardless of the interpolation scheme used.

Identifying actual data for the data base is not always easy or straightforward. Typical proposed waterway modifications usually involve some widening and deepening of existing channels or perhaps changing the channel path. Some projects involve dredging channels where none had existed before. In cases where the channel dimensions of a new design are specified, the bathymetry can be read directly from the plans for the waterway.

Much of the overall project area may be in a natural state or may be the result of previous dredging. Many available charts of waterways are not recent, and few of these include information on water depth that is dense enough for an adequate data base. Most field survey records provide discrete soundings at specific data points rather than a continuous bottom profile. About 60 percent of field surveys conducted by the National Oceanographic and Atmospheric Administration (NOAA) were done prior to 1940 with lead lines (NOAA, unpublished data).

Waterways are not static; they are constantly changing. Some bathymetric changes are due to seasonal variations of flow, others may be part of variations resulting from singular events that occur every few years (for example, floods), and still others represent long-term trends that may span decades, if not centuries. Investigation and correction of chart discrepancies reported by various sources are backlogged, with about 20 thousand discrepancies remaining unresolved in backlog during early 1991. NOAA can field investigate about 20 percent of chart corrections, which leaves major areas with unresolved discrepancies. As a result, reliable continuous bottom profiles are available for only some of the important shipping routes along the coasts and in ports and waterways (NOAA, unpublished data). Therefore, developing a bathymetric data base requires careful research and may well require the supplementation of information on available charts with in situ measurements. It should be noted that the density of bathymetry data points required for determining channel flow and grounding is more demanding than that required for determining of the forces on a ship (Norrbin, 1978; Norrbin et al., 1978).

WATERWAY ENVIRONMENT

Because of the efficiency that results in the computer programming, the data base framework selected for the waterway environment is usually identical, or at least corresponds quite closely, to that for the waterway bathymetry. In this way, similar interpolation schemes can be used for both. However, determining the waterway environment data base is fundamentally more complicated than for waterway bathymetry. Those quantities that describe the environment, such as wind, current, and density, often vary with time of day or season or with altitude or depth in the waterway. For existing waterways, information on existing charts regarding currents is typically even less detailed than that for bathymetry, and information on other quantities is even more sketchy. For waterway designs involving changes in existing bathymetry, information on current variations needs to be developed.

The waterway environment can reflect some unique problems. In some cases, a density stratification may exist (for instance, at a river mouth where fresh water may override a saltwater wedge). In such cases, the variation of current with water depth can even include a reversal of the flow. Similarly, air characteristics, such as velocity, turbulence, and temperature, can vary with weather or with altitude above the waterway and can be significantly different in the shadow of buildings or bridges than elsewhere. Design-related bathymetric changes relative to the tidal prism in coastal ports may also affect sedimentation rates and, consequently, waterway operations and maintenance. Data on such effects are generally not available, but depths could be changed in the simulation to obtain a rough estimate of behavioral changes in the design ship when sedimentation modifies the bottom profile. However, there is no indication that maintenance factors have been incorporated into most simulations.

The database for the environment can be formed in several ways. For an existing waterway, a field survey can be conducted to determine the values in situ, but the cost of such a survey may be high. Hydraulically scaled models are traditionally used either as a less-expensive alternative to in situ measurements in existing waterways or as a way to determine the flow in waterways not yet built. These models usually predict reliably the gross characteristics of horizontal flow. However, due to difficulties in scaling viscous effects, predictions of vertical variation of fluid velocity at any given point are less reliable.

Computational fluid dynamics (CFD) schemes have been developed in the past decade to predict currents in waterways with complicated bathymetry. Already these methods are less expensive to use than physical models. As with physical models, CFD schemes yield better results for the average horizontal fluid velocity than they do for the vertical fluid velocity distribution at a given point. However, both hydraulic models and CFD schemes can benefit from comparison with in situ measurements.

It is very difficult to determine the variations of the waterway environment that occur with depth or altitude. More importantly, no validated means exist for predicting the effect of these variations on the forces acting on the ship. Therefore, it is typical to replace the variation of current with depth or the variation of wind velocity with altitude by a single, uniform current or wind vector that will produce approximately the same force distribution on the vessel. In this case, the actual value of the current that is not depth dependent may be entered into the data base and is chosen carefully to reflect the more complicated character of the actual flow. In particular, the value appropriate for one ship loading and draft may not be appropriate for the same ship at a different loading and draft.

Some facilities retain the vertical variation of the current with depth in their data bases and estimate the effective value of the current as a value of current averaged over the actual ship draft at the given location. This scheme requires a much bigger data base and more computation, but it has the advantage of not requiring revision if a different ship or ship loading is used for the simulation. Finally, there is usually not one but a collection of environmental data bases, each reflecting a given state (phase of the tide, current distribution, and weather).

MATHEMATICAL MODEL OF SHIP DYNAMICS

The framework for the theoretical model of ship dynamics was described in general terms in Chapter 3. It involves two separate pieces: Newton's equations of motion (as modified by Euler for moving bodies) and a representation of the forces acting on the ship as a function of its orientation in the waterway and with respect to environmental conditions. The Euler equations of motion have a sound scientific base. The coefficients associated with these equations are easily identified and are therefore not discussed further in this report. Essentially there is no variation in this part of the framework from one facility to another.

Because Euler's equations are not in question, the accuracy of the mathematical model of ship dynamics is governed by the ability to predict the instantaneous force system on the ship. (For brevity of discussion, this report does not distinguish between forces and moments, referring to both simply as *forces*). The forces acting on the ship arise primarily from the combined effects of water surrounding the ship, wind, waterway geometry, and other external forces such as tug boat assistance and riding on anchor (Abkowitz, 1964; Bernitsas and Kekridis, 1985; Eda and Crane, 1965; Norrbin, 1970). Most of the complexity (and uncertainty) of a mathematical model for the behavior of a ship stems from the estimates made for this force system. Considerable variation exists from one facility to another because representations of the forces that act on the ship are complicated and do not have the firm scientific basis of Euler's equations.

The dynamic framework is usually separated into several manageable constituent parts (or modules), which are dealt with relatively independently as shown in Figure 5-1. The separation of the ship hydrodynamic forces into those in unrestricted shallow water and corrections to account for restrictions, such as banks, reflects the historical development of mathematical modeling of maneuvering ships over the last 100 years. Figure 5-1 depicts three threads of information (represented as thick horizontal lines) that affect several modules within the ship model. Two of these, the commands from the pilot and the position and velocities resulting from the ship's behavior, are available outside the ship model. The third, which is the sum of instantaneous forces on the ship, is part of the necessary internal bookkeeping for computing the ship's motion.

In many simulators, only three degrees of freedom are used (surge, sway, and yaw—the so-called horizontal motions) because the vertical motions interact little with the steering and maneuvering characteristics of the ship. In a severe turn, the ship roll angle may become large for ships with small inherent roll stability. The angle of roll changes the wetted hull shape. This can substantially increase the turn radius. Where the under-keel clearance is small, the vertical motions (heave, pitch, and roll) can

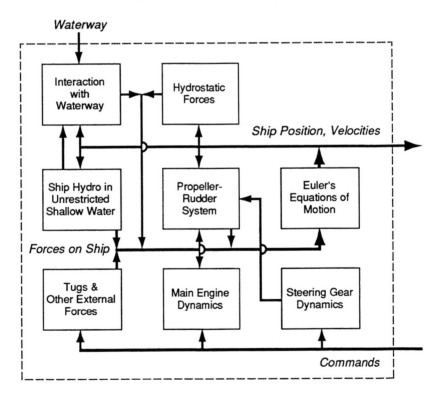

FIGURE 5-1   Schematic diagram of modules in simulated ship behavior.

have an important effect due to the combined effects of squat and the response to waves, currents, or wind. In these circumstances, all six degrees of freedom are used.

COMPONENTS OF THE FORCES SYSTEM

In the following sections, various components of the forces system acting on the ship are discussed in general terms, including approximations used for application in a simulator and the identification of numerical parameters. Characterization of the hydrodynamic forces on the ship is usually treated as a variation and expansion of the classical treatment of steering and maneuvering in deep water. Therefore, the deepwater problem is discussed first, even though it is not applicable to typical waterway design. Components are also discussed in relation to unrestricted shallow water, restricted shallow water, rudder-propeller systems, and propulsion and steering systems.

Specific equations are not introduced in the following sections. The mathematical presentation of any of these models is algebraically intensive, as demonstrated by a mathematical model for the *Esso Osaka* in unrestricted shallow water (for further information on *Esso Osaka,* see Abkowitz, 1984; Ankudinov and Miller, 1977; Crane, 1979a,b; Dand and Hood, 1983; Eda, 1979b; Fujino, 1982; Gronarz, 1988; Miller, 1980; Report of the Maneuvering Committee, 1987).

Deepwater Factors

Measurement of the steering and maneuvering characteristics of ships in deep water is a well-understood and highly developed technology. Most facilities use a history-independent formulation where the forces are assumed to be approximately the same as those that would exist on a ship that has been in the same situation for a long time. Forces acting on the ship are assumed to depend only on the instantaneous attitude velocities and accelerations of the ship (referred to simply as the instantaneous *state* of the ship). It is assumed that these forces do not depend on the motions of the ship or its attitude at previous times. Indeed, *memory effects* are a well-known phenomenon resulting from the wave system and viscous flow created by the ship's forward way and by wave-induced motions, and these effects are important in predicting the oscillatory motions of a ship due to a seaway. However, time scales for the steering and maneuvering problem are so large that these memory effects are unimportant in this context.

The framework usually consists of a polynomial representation of the forces in terms of the instantaneous displacements, velocities and accelerations of the ship, propeller, and rudder (and various products of these motions). This polynomial can be viewed as a truncated, multivariate Taylor's expansion about the state of the ship, which corresponds to straight-line travel at a constant forward speed. This representation does not embody any physics per se, but simply reflects an implicit assumption that these forces vary smoothly with the state of the ship. The expansion is truncated to include only those higher-order terms that appear to yield significant forces. Quantification of the framework is obtained by identifying the coefficients of each term in this polynomial. In fact, many different mathematical frameworks are used at simulation facilities around the world, and each facility appears to have its favorite. Most of these frameworks are identical in their linear terms and in many of their nonlinear terms. Differences occur in the number and type of higher-order (that is, nonlinear) terms that are retained. However, it should be noted that the numerical values of the coefficients associated with the linear terms depend on which nonlinear terms are retained in the framework.

The coefficients that relate instantaneous motions to forces acting on the ship are most often determined experimentally by captive model tests using either an apparatus called the Planar Motion Mechanism (PMM) or a special facility called a rotating arm basin. These tests are performed by oscillating a laterally restrained scale model of the ship in question in sway and yaw at Froude-scaled test conditions. It is assumed that viscous effects (which are not scaled in the model tests) can either be ignored or corrected for. Analysis of the time histories of the forces acting on the ship model resulting from many captive model tests is used to determine both linear and nonlinear coefficients in a mathematical model for these terms. These coefficients are obtained by a multivariate regression or by curve fitting, depending on the conduct of the captive model test. In addition, tests are performed with the rudder at various angles and the propeller at various rotational speeds. Changes in forces and moments resulting therefrom are also identified by coefficients in polynomial framework.

The mathematical model for hydrodynamic forces and moments is joined with Euler's equations of motion and a model for the dynamics of the propulsion system (discussed separately below) to form a simulation model for deep water. This model can be used for simulating steering and maneuvering exercises in deep water and for training of a ship's bridge team. Such models have been used by Japanese shipbuilders, for instance, to select the size and location of rudders in new tanker designs.

Because captive model tests are expensive and time consuming, many facilities have built up libraries of dynamic data on previously tested models. These data have been used by some of these facilities as a data base from which the coefficients in the mathematical model for ships can be estimated by regression (that is, without a physical model test). Presumably, if the data base were large enough, this approach would be successful.

However, most facilities do not release their data, and thus, it is difficult to judge the success of this process.

In recent years, an alternative scheme called *systems identification* has been devised for determining the coefficients in the mathematical framework for all the hydrodynamic forces, including the propeller and rudder (Abkowitz, 1980; Aström et al., 1975). In this scheme, a free-running model (or full-scale ship) is instrumented to record both the motions and inputs (for example, rudder angle, propeller revolutions per minute [RPM], speed, heading). This information together with a proposed framework is used to "identify" the numerical value of the coefficients and to give a measure of "goodness of fit." The mathematics are too involved to attempt to describe in this report. If these data are taken on a model, then some correction for the viscous effects may be called for; if these data are taken on the full-scale ship, then the coefficients may be used directly. Some indications suggest that this approach can be as successful as using captive model tests, although the systems identification approach typically identifies fewer coefficients than are used in the traditional approach.

Interestingly, neither analytical hydrodynamic analysis nor computer-based algorithms (CFD codes) are sufficiently mature to predict coefficients for use in steering and maneuvering models from the underwater geometry of the ship, even for this simplest case of deep water. The difficulty lies in the fact that viscosity has important effects and cannot be ignored. Advances are being made in developing computer-based programs for treating viscous free-surface flows. However, these programs may be as expensive to run as physical model tests, and their ability to reproduce physical model test results has not been demonstrated.

The simulation of steering and maneuvering in deep water appears to be satisfactory for engineering applications, as long as the coefficients of the mathematical model are identified by a properly conducted physical model test. Using a data base of test results to predict the coefficients of a ship without a model test may be acceptable for most waterway work (Clarke, 1972; Kijima et al., 1990).

Unrestricted Shallow Water

The maneuvering of ships in unrestricted shallow water (water of a depth less than 2.5 times the vessel draft of infinite lateral extent) has been investigated much less than that of deep water. The flow around a ship becomes dependent on the water depth, and this additional parameter makes both theoretical developments and experiments much more difficult. Nonetheless, nothing about these experiments makes the interpretation of the results more complicated or more difficult than the deepwater case, except in the instance of extremely shallow water where the viscous flow under the ship's bottom may not be modeled well in small-scale experiments. In particular, the same mathematical framework typically is adopted for the force model, with perhaps a few more nonlinear terms included to capture forces that are important in shallow water but are inconsequential in deep water.

Several experimental studies have been performed in moderately shallow water, and their results are surprising. Whereas the force coefficients in the mathematical framework vary smoothly with water depth, some of the handling characteristics do not. For instance, several researchers using model tests found that ship turning performance first improves upon entering shallow water and then degrades rapidly as the under-keel clearance becomes very small (Crane, 1979a; Fujino, 1968, 1970). This finding suggests that the effect of very low under-keel clearance can be dramatic and cannot be ignored. No measurements, full or model scale, have been made in the range of 10 percent under-keel clearance or less, a range commonplace in U.S. ports (National Research Council, 1985).

To obtain experimental data for use in the mathematical framework, it is necessary to run the same type of PMM tests or systems identification study for deepwater cases, but at several finite water depths as well. This approach requires a test basin where the bottom is extremely flat; few such basins exist worldwide. As a result, very few ship models have actually gone through extensive shallow water maneuvering testing, and the data are sparse. Available data have been referred to extensively.

The situation in unrestricted shallow water is similar to that in deep water. However, not all the phenomena are clear. To perform either physical model tests or full-scale trials would require addressing significant modeling questions concerning the viscous flow in the gap between the ship and bottom and concerning the deformation of the mud bottom by the ship. The cost of performing the required tests is high because a new test parameter (water depth) must be varied. The lack of a flat bottom at most facilities has inhibited the testing of ship models with under-keel clearances comparable to current ship traffic. With the help of some theoretical developments, most ship model testing facilities have developed proprietary, semiheuristic schemes to modify deepwater maneuvering coefficients so that they are approximately correct for shallow water.

Restricted Shallow Water

The preceding discussion of the ship model focused on maneuvering a ship in unrestricted, quiescent water of finite depth. However, many other interactions need to be considered if the simulator is to be useful in waterway design. Interactions include the force system on a ship maneuvering in a channel with geometric complexity (turns, banks, uneven bottom, and so on), with hydrodynamic complexity (complex current patterns, tidal variations, and so on), and with atmospheric disturbances. It is convenient to separate differences in these areas into two force systems: one resulting from the atmospheric environment and the other resulting from the water environment. Interactions due to other vessel traffic and the use of auxiliary help, such as tug boats, also need to be considered.

The effect of wind, resulting from both the average velocity and gusts, can be important in some waterways. Wind forces become relatively more important when the vessel has small forward movement, when the vessel has a large "sail area," or when it has a shallow draft. Sail area is affected by hull and superstructure configurations, freeboard, and deck cargo such as containers. With loading, the sail area of a tanker decreases, and its draft increases, making a fully loaded tanker less susceptible to wind effects. A containership loaded with empty containers that are stacked high on deck may have both a large sail area and a small draft, and thus it is very vulnerable to wind effects. When the wind is parallel to the channel and in the same direction of travel as the ship, controlling the forward movement can be difficult, especially for diesel-powered ships where the minimum sustainable RPM corresponds to a significant speed and where the number of air starts may be limited.

Significant wind forces usually arise when the wind velocity is much greater than the ship velocity, and as a result, a simple framework for these forces is usually adopted. Aerodynamic forces are estimated using an empirical drag coefficient dependent on the relative wind direction. The effects of gusty conditions are usually included as an increment to the average wind velocity.

The framework for the hydrodynamic forces is a set of equations used to predict the *changes* between the force system resulting from these interactions and the force system that would exist in unrestricted shallow water of the same depth. This framework usually has the same general polynomial format as that used for hydrodynamic forces in unrestricted shallow water. The coefficients now depend, however, on the distance to, and the character of, the bank and other obstacles.

This force system consists of steady forces and unsteady forces. Steady forces are typically due to an interaction with a bank. When the ship is travelling parallel to the bank, force is directed toward the bank (so-called bank suction forces), and the moment results in a bow out movement. However, at other angles, changes in these forces can be either toward or away from the bank. Propeller revolutions can also affect these forces in the presence of a bank. A considerable body of literature on these steady forces exists where the results of experiments are reported (Norrbin, 1970, 1978). Empirical formulas have been developed that are successful for predicting them.

Unsteady forces are usually separated into two types. The first or quasisteady force system represents a modification of the steady force system due to the instantaneous motions of the ship due to the proximity of restrictions. The second or fundamentally unsteady force system represents the transient forces that result from the ship approaching a bank or obstruction, passing by a discontinuous bank, passing another ship (either on reciprocal courses or overtaking), or passing into an area where the water depth changes suddenly or the water current varies dramatically in speed or direction (see Armstrong, 1980; Crenshaw, 1975; Plummer, 1966).

The quasisteady force system arises when the ship is traveling, on the average, parallel to a continuous bank of uniform geometry in a region and where the depth changes are very gradual and the current is nearly constant in speed and direction. In this case, motions arising from course keeping can be considered as small perturbations about an otherwise steady flow. The quasisteady force system is usually characterized by the same framework as that for unrestricted shallow waters, except that the coefficients must include an additional parameter: the distance from the bank. Coefficients in this framework depend not only on water depth and ship geometry, but on current in the waterway and geometry of the bank as well (Abkowitz, 1964). For this situation, it is also possible to perform PMM testing at several different water depths and, at each of these depths, perform additional testing at several different distances from the bank. However, the number of variables involved make the cost of this type of model test program high. Thus, such tests are almost never conducted to identify these coefficients. Nevertheless, some tests of this type have been performed, and results are available in the literature (Abkowitz, 1980; Eda et al., 1986; Norrbin, 1978).

When the ship is not traveling approximately parallel to the channel or is oriented to other traffic so that the flow is fundamentally unsteady, it is impossible either to eliminate time (that is, history) from the problem or to reduce the transient force system to simple time-independent coefficients. The most studied of these fundamentally unsteady phenomena are cases of ships passing interrupted banks, ships approaching banks, and ships passing one another (Dand, 1984). The literature in this area is very limited, and most of the data that are available are for the passing ship case.

Experimental studies have been conducted on the effect of interrupted bank systems where the interruptions are in a straight line (Norrbin, 1974, 1978). Reducing these data to numerical formulas appears to have been accomplished by various facilities using proprietary techniques. The effect on the force system due to sudden changes in waterway depth, to a waterway bathymetry that is truly three-dimensional, or to currents that vary significantly along the length of the ship apparently have not been systematically studied. However, mathematical simulation models typically ignore or only crudely approximate the effects from this kind of temporal or spatial dependence. The computation of this representative value from the instantaneous state of the ship and its position in the waterway is heuristic and varies considerably from facility to facility.

Model tests to determine the force and moment history of two ships passing one another have been conducted in several contexts and particularly for the Panama Canal study (see Appendix C). The overtaking configuration is, in general, the most severe because the time during which the interaction between vessels may be strong is far longer, although studies of meeting situations are more common. Interaction forces between the two hulls will cause perturbations in the trajectory of both ships, particularly if the waterway is narrow (Gates, 1989; Hooyer, 1983; Plummer, 1966).

Potential parameters in such a study are numerous and include the description of the two ships, each of their speeds, initial passing distance, passing angle, water depth, and distance to a bank. Parametric tests to investigate each of these variables appears feasible, but such tests probably would be prohibitively expensive. The usual practice (when passing tests are conducted at all) is to measure the force system when passing ships are constrained to straight-line motion. Fundamentally unsteady forces and moments are measured, but deviations of the ships' tracks in response to these forces are not allowed. These responses may be significant, especially when the passage is a close one or when ships are in an overtaking configuration (where the exposure time is long). Typically, constrained model test data are used, together with empirical or heuristic corrections, to predict the force and moment history for the actual passing condition.

A body of theoretical literature also exists based on a linear (small motion) analysis of a ship passing a bank or other objects (Yeung, 1978). These theoretical developments often are used to establish framework elements of the unsteady waterway interaction framework. Coefficients associated with this framework are usually identified using the above-mentioned experimental results available in the literature, modified to account for differences between the ship under consideration and the ship that was tested. These semiheuristic methods are almost always proprietary to the individual facility.

Finally, there are other possible important interactions that may be required for certain simulations. Tug boat assistance is a feature of many maneuvering situations. The presence of tugs alongside a larger ship is, like the passage of ships, a situation where a strong interaction is expected in principle. However, because these tugs are typically much smaller than the simulated ship, their principal interaction is through the thrust (both size and direction) generated by the propeller-rudder combination (Brady, 1967; Dand, 1975; Reid, 1975, 1986). In general, this interaction is directed by the pilot or master of the simulated ship, and the modeling of this interaction is typically treated in a quite simple fashion.

Rudder-Propeller System

This force system module represents the combined effects of the propeller and rudder, which are usually treated together because they are the primary actuators for steering and maneuvering. Rudder angle, propeller RPM, and propeller pitch (if the propeller is variable pitch) are introduced as new variables, and the forces resulting from the interaction of propeller, rudder, and hull typically are characterized by them. Because these forces also depend strongly on the flow about the basic ship, formulas for these forces also involve the state of the ship and its geometry (particularly the after body).

The force and flow field produced by a propeller driving a ship at constant speed are relatively well known, and means for its prediction are available. The force and flow field created by a propeller spinning at a speed different from these equilibrium conditions is less well known, especially when the ship is maneuvering and the propeller may be spinning with a rotation that would ultimately cause the ship to reverse its present direction. Four separate situations with regard to propeller operation can be identified, depending on the sign of the velocity of the ship (either ahead or astern) and the sign of the propeller rotation (either in the ahead direction or the astern direction). These four situations are usually called quadrants, because they appear on a graph of ship speed along one axis, and propeller RPM appears along the other. Characterizing the effect of the propeller for all possibilities of ahead and reverse propeller rotation, and forward and astern ship's velocity (the so-called four quadrant problem) is difficult. Most simulators do, however, include an approximate model for these conditions.

The side forces on a rudder are usually proportional to rudder angle when small rudder angles are used, but depend in a more nonlinear fashion for large rudder angles. Side forces on a rudder also depend approximately quadratically on the flow velocity over the rudder, and thus, the hydrodynamic effects of the propeller and rudder are fundamentally linked. When the ship is proceeding ahead and the propeller is rotating to maintain this motion, flow over the rudder is typically at a somewhat higher velocity than the ship's velocity. However, if the pilot decides to execute a full-astern maneuver (or the pitch of the propeller is reversed), then flow through the propeller is ultimately reversed, and the rudder may experience little or no flow over it. This situation is often referred to as *blanketing the rudder* and results in the rudder being almost ineffective. A characterization of these effects using elementary hydrodynamic analysis and empirical results is usually included in a semiempirical model for the propeller-rudder system.

Various facilities differ in their approach to quantifying propeller-rudder interactions. Because a Froude-scaled ship model does not reproduce the viscous effects properly, a self-propelled ship model cannot behave as the full-scale ship would. That is, the propeller in a self-propelled ship model has to produce considerably more thrust to overcome the relatively greater viscous drag of the model. The propeller-rudder interaction forces are often measured on a captive, towed model with the propeller spinning at a range of RPMs and at various rudder angles. The results are scaled up to full scale using the information from separate propeller tests using larger models, performed in a facility called a propeller tunnel. This type of facility models the atmosphere so that important effects of cavitation can also be modeled and observed.

The cost of experimentally determining influences of the propeller and rudder is high. Many facilities use empirical formulas based on previous model tests to estimate the four-quadrant behavior of the propeller and its interaction with the rudder.

Additional modules are often added to account for other maneuvering devices, such as thrusters, if they are installed. Characterizing these devices and their interaction with the hull is in principal very complicated. As a result, a semiempirical approach is usually adopted.

Model of Propulsion and Steering Systems

The propulsion and steering systems are also critical to maneuvering a ship, because the propeller RPM and rudder angle are determined by them. They are also mechanical devices with their own dynamics. These devices cannot respond instantly when commanded because of their own inertias and other limitations. A detailed characterization of these maneuvering elements would involve developing equations of motion that reflect the physical properties or response of many individual components. Steering gears and thrusters have relatively straightforward mechanisms, and they apparently do not require great sophistication in the mathematical model to capture their behavior.

Characterizing the main propulsion system behavior is, however, more difficult because typical systems are large, have substantial inertias, and involve many components, particularly for diesel systems. The propulsion model (usually referred to as the engine model) also requires characterization of the torque characteristics of the propeller as a function of its RPM. Two choices are typical for main propulsion: steam turbines and diesel engines.

Steam turbines have few moving parts in the main drive train to model. These include the rotary inertia and friction of the turbine rotors, gear system, line shafting, and propeller. Because these elements are geared together, they are dynamically equivalent to a single rotating mass. These characteristics result from the thrust the propeller produces and its hydrodynamic losses. In addition, dynamics involving the steam valves and associated equipment may be important. Models for complete steam turbine power plants are somewhat complex, but reliable models have been constructed by several different facilities (van Berlekom and Goddard, 1972).

In today's fleet of merchant ships, diesel engines are much more popular choices for the main propulsion plant and are, unfortunately, much more difficult to characterize. Large, direct-connected diesel engines typically have 6 to 12 cylinders and are equipped with many auxiliary mechanical components, such as turbosuperchargers. The sheer number of moving parts in such an engine and the associated degrees of freedom preclude direct modeling of the intercoupled mechanics of each component. Rather, an indirect, behavioral model is usually adopted, where the engine in toto is replaced by an equivalent dynamic system with only a few degrees of freedom and with inertias and damping chosen to mimic the behavior of the diesel engine.

In addition to the mechanical modeling of the main elements of a diesel engine, other modeling problems exist. Starting and reversing these machines are achieved by injecting compressed air into some of the cylinders. Although this process is fairly reliable, failure to restart is not uncommon, especially in cold weather. Thus, a random delay may occur in the reversal of the engine. Further, some diesel engines have a finite reserve of starting air, and the reversal-restart cycle may become compromised if many such maneuvers must be performed in close succession. During changes in power level for some configurations of diesel engines, a significant lag may also occur in the air boost pressure due to the dynamics of the turbosupercharger-air plenum system. Thus, modeling the dynamic performance of a diesel engine during maneuvering is a significantly greater challenge than modeling a steam turbine, and the state of the art is not as well developed (Eskola, 1986).

SUMMARY

The mathematical model used for shiphandling simulation consists of not one model, but a series of many models, each representing a particular piece of hardware or important physics. These models are interconnected inside the computer that runs the simulator to reflect the physical interactions among the elements they represent. Each of these component mathematical models has its own set of uncertainties resulting from the modeling process, and it is difficult to assign an uncertainty for the overall model. The model that predicts the hydrodynamic forces on a ship as a result of its motions and proximity to the bottom, banks, and other waterway features is perhaps the most difficult to develop, and its uncertainty is greatest in the case of shallow, restricted channels.

6

Assessment of Simulator Technology and Results

As described in the previous two sections, simulation involves an array of physical and mathematical components, each with their own limitations and inaccuracies. In addition, real-time simulation uses a pilot who introduces human variations into the simulation. If the results of the simulation are to be interpreted sensibly in the waterway design process, it is important to determine how well this array will predict the track of a ship in a given situation. Discussion in the preceding chapters shows that it is difficult, if not impossible, to treat the question quantitatively and scientifically. With this caveat, the following discussion assesses simulator technology from an engineering point of view, that is, in the context of its application to waterway design.

ACCURACY

Simulation has only recently become a feature of some waterway design initiatives, although use of the technology is increasing. Interpretation of the simulated vessel tracks provides insight into the various navigation factors (principally turn characteristics, channel width, and depth). The assessment presented here addresses the related concepts of accuracy and validity of simulation in the context of the waterway design process. For this discussion, a simulation will be considered accurate if it can produce piloted track predictions that are useful as a basis for a design decision concerning navigation and risk. Accepted guidelines for this accuracy apparently do not exist, and the accuracy requirement varies depending on the exact nature of the design problem. Construction tolerance for horizontal dimensions of the waterway is about 10 feet for fairways. Therefore, comparable accuracy for simulation is a reasonable goal for channels or approaches to berths involving dredging; greater accuracy is only of academic interest. However, for berthing or lock operations with very tight maneuvering tolerances, the level of precision required for the simulation is correspondingly higher.

Validity of a simulation can be expressed in the form of two narrow but scientific assessments. First, does the predicted track of a given ship accurately reproduce the real ship track when the pilot or autopilot performs at the simulator exactly as either would perform on a real ship? More scientifically, is the output of the simulator the same as that of the ship when the input to both is identical. If the simulator meets this criterion, then, will the pilot (or autopilot) make the same maneuvers at the same times in the simulator environment as would be made in the shipboard environment given the same transit conditions?

If the answers to both questions are affirmative, then the simulation can clearly be considered valid. That is, the predicted tracks will compare well with the full-scale results for a piloted ship. For discussion purposes, the first of these is identified as the mathematical modeling problem and the second as the pilotage modeling problem. In the past, these two very different aspects of model validation have often been intermingled.

Assessing the validity of a simulation in terms of the separate accuracies of the mathematical model and of the pilot model is used here for convenience of discussion. The mathematical modeling described above is the *open loop* response of the ship (that is, without the use of corrective steering measures). A mathematical model that is accurate by these terms can perform *dead reckoning*, a computation of the track, given only the history of the commands. This significantly more sensitive problem poses a particularly severe demand on accuracy.

In either the full-scale or simulated transit, the pilot takes corrective action when the ship appears to deviate from the strategy for transit, no matter what the cause. Thus, the pilot in a simulator will attempt to correct any deviation from the planned track resulting from an error in the mathematical model, just as if some real deviation was caused by the proximity of a bank, vessel, or other waterway feature. As a result, the tracks of all types of ships tend to be close to one another, independent of their inherent maneuvering behavior. The pilot's skill enables anticipation of the ship's behavior and its interaction with the environment so that commands are given expeditiously. This result is achieved, even for ships that are difficult to steer. Thus, the errors in the mathematical modeling are particularly difficult to discover from an analysis of piloted tracks. On the other hand, the accuracy demand for the mathematical model used in a shiphandling simulation can be less than that required of dead-reckoning simulation. That is, feedback provided by the pilot may minimize the effects of inaccuracy. During this study, no research was identified that can provide guidance for relating these two accuracies. In the sections below, a critique is presented for the state of practice of the two aspects: the mathematical modeling problem and the pilotage modeling problem. These discussions, however, remain rather general because, as the above discussion reveals, the accuracy required of these components cannot be specified within the current state of practice.

CRITIQUE OF MATHEMATICAL MODELING TECHNOLOGY

The discussion in Chapter 5 demonstrates that many acceptable data base frameworks for the waterway and its environment are possible. Identification of the constants for the data base is direct for bathymetry, but may require either physical or computational fluid dynamics (CFD) modeling for the currents or winds. Thus, tools are available for developing these data bases, but the cost of determining the appropriate input data and of exercising these tools limits their use.

Mathematical models for ship behavior are well known in the case of steering and maneuvering in deep water. The measurement of turning performance in deep water is usually included in the trials of new ships. As a result, many comparisons with actual ship data have been made between simulator predictions for deep-water maneuvering and full-scale gross measures of maneuvering performance (for example, advance, transfer, tactical diameter, directional stability). The state of practice is such that theoretical predictions of deep-water turning performance are typically within 10 to 20 percent for these measures when the coefficients in the particular mathematical framework are identified using scaled physical experiments (captive model tests and extensive propeller-rudder interaction model tests).

Unfortunately, the performance of physical model maneuvering tests on new ship designs is not common. The identification of coefficients in the mathematical framework for new ships is often performed by interpolating within a data base of coefficients for similar ships (that is, without using physical experiments). This approach appears to be successful if the new ship is indeed similar to those in the data base, and the degree of success depends critically on both the size and quality of the database and on a careful review of the resulting coefficients by a knowledgeable practitioner.

The frameworks and coefficient identification process in use for both unrestricted and restricted shallow water vary from simulation facility to simulation facility. Almost all of these mathematical models are considered proprietary to the individual facilities and were not available for scrutiny or detailed comments in this report. As discussed in Chapter 5, the technology to identify coefficients in any such framework is available if the under-keel clearance is greater than 10 percent. Only a small part of this technology is ever involved in any simulation because of the extremely high cost of the required experimental setup and of the data collection. For very small under-keel clearances, the physical phenomena remain unclear.

In the current state of practice, many frameworks and associated coefficients for models that deal with details specific to waterway design (shallow water, banks, passing ships, variable currents, and so on) are constructed heuristically by using an amalgam of available theoretical developments, by using results of the limited available model tests of ship performance in idealized waterway configurations, and by bold assumption. In addition, there appears to be little scientific basis for the usual, quasi-steady treatment of highly time-dependent events, yet these are critical situations in many simulations.

Scientific verification of the accuracy of available models by comparison with full-scale results generally is also missing. Full-scale measurement of maneuvering tracks in unrestricted shallow water is limited. The most extensive set of tests appears to be those conducted on the *Esso Osaka* 10 years ago (Abkowitz, 1984; Ankudinov and Miller, 1977; Bogdonov et al., 1987; Dand and Hood, 1983; Eda, 1979b; Fujino, 1982; Gronarz, 1988; Miller, 1980). Even so, these tests were performed at under-keel clearances that are still large compared to those tolerated in many waterways.

Thus, the parts of the mathematical modeling process that are critical to simulation results but lack scientific precision are precisely those aspects that differentiate simulation for waterway design from simulation for deepwater maneuvering. These include:

• modeling of the hydrodynamic forces and moments for situations where the under-keel clearance is small;

• determination of the forces on a ship passing near waterway sides (banks);

• determination of the forces on a ship in essentially unsteady conditions (approaching banks, approaching and passing other ships, moving into regions with sharp current or bathymetric gradients); and

• the implicit assumption in most frameworks that the forces resulting from the various phenomena (for example, bank effects, propulsion, rudder effects) can be superimposed without considering their interaction.

Because of all the additional assumptions, it is unreasonable to expect that state-of-the-art mathematical models for maneuvering in restricted shallow water will be as accurate as those for deepwater maneuvering. However, as discussed above, the accuracy required for the mathematical model (open loop) may be considerably less than the accuracy needed for the waterway design (closed-loop). This consideration, together with the significant costs involved, has inhibited the development of more accurate mathematical models.

In practice, the validity of the mathematical model is established by comparing it with one or more of the following:

- real-world measurements or data, such as ship test trials;
- results from tests conducted using measurements derived from scale-model tank testing;
- performance estimates derived through mathematical extrapolation or interpolation using accepted theoretical models;
- the performance expected and evaluated by experts on the system the simulation has been designed to model; and
- the performance expected and evaluated by an interdisciplinary team of participants in the design process, including sponsors, planners, designers, and pilots, for the system the simulation has been designed to model.

CRITIQUE OF PILOTAGE MODELING TECHNOLOGY

A properly developed mathematical model will predict with acceptable accuracy the motions performance of a ship plying a waterway in response to commands from its pilot. The pilotage problem is this: Will the pilot perform on the simulator as if on board an actual ship? This question is much different from, and more difficult than, the mathematical modeling aspect because it is, to a major extent, a physiological and psychological question. If the pilot is a human pilot, then one can anticipate that slightly different commands will be given during each transit, even if the conditions during the passage are exactly the same. Thus, the sets of commands for a series of like passages by a single human pilot will be only approximately alike. Variations of this type are not encountered in standard fast-time simulation. If wide variations in shiphandling do result from man-in-the-loop simulations, the implications for waterway design may be important. If it can be determined that a particular waterway configuration is highly sensitive to pilot performance, then it would be prudent to search for and consider an alignment that is less sensitive: for example, a more desirable alternative would show little variation in swept path amongst different pilots.

The bridge, visual scene, and radar contribute to an observer's judgment of the *face validity* (also referred to as apparent validity) or realism of a simulation. A full-size ship's bridge, a high-fidelity visual scene, and a "stimulated" real radar set have high face validity. In turn, such a judgment contributes to the observer's acceptance of the simulator, design study, and eventual implementation of the findings. Physical surroundings may also contribute indirectly to pilot-ship-environment system performance by affecting the shiphandler's motivation.

Thus, a relationship exists between fidelity of the simulator and pilotage accuracy, but the relationship is difficult to pin down. The omission of real-world elements may influence the pilot to make a command that would not be made on board ship, or vice versa. Such an omission may be a major element (for example, the absence of a compass repeater in a place where it can easily be referred to) or the omission of a seemingly small item (for example, a distant church steeple that the pilot normally uses as a navigational reference). Because it is difficult to know a priori what is important to portray in a simulation and what is not, designers and pilots generally have more confidence in a high-fidelity, real-time simulator than in a low-fidelity one.

Whether or not fidelity adds to the accuracy of the pilotage modeling, it certainly adds considerably to the cost of simulation. For example, including a prominent water tower next to the church steeple in the visual scene mentioned above may only contribute to face validity, the immediate impression of realism. However, face validity can contribute to user acceptance of the simulator and its results. If the steeple is not used by local pilots as a visual cue in the piloting process, however, its inclusion may add little or nothing to the accuracy or fidelity of the simulated pilotage.

The question of how much fidelity is needed to achieve accurate simulation is different from how much fidelity is needed to make the pilot perform realistically in the simulator. Pilots, by the nature of their profession, need to be quick learners and exceedingly adaptable. For instance, pilot skill includes the safe piloting of vessels from the pilot's very first pilotage on vessels of that class. This flexibility helps pilots interpret and use modest simulation that can only be called low fidelity. A typical low-fidelity simulator may have, for instance, only one crude television-sized display that can be switched from a synthetic radar view to a low-resolution, dead-ahead view. In these situations, pilots with experience in operating ships and knowledge of the waterway in question may be able to fill in the missing information and produce a track similar to that achieved on a high-fidelity simulator. Moreover, it is possible, even in a low-level simulation, to provide the pilot with a much more accurate view of the situation than will be available on board ship (for instance, with an accurate bird's-eye view). In this case, the results of the simulation may underplay the safety factors (Perdok and Elzinga, 1984; Schuffel, 1984). Because a high-fidelity simulation can be quite costly, the demonstration of validity and user acceptance for a low-fidelity simulation could lead to increased use of the latter.

The accuracy of pilotage modeling in a real-time simulator is much more difficult to ascertain than that with mathematical modeling of the basic ship behavior. No objective measure was found for it during the study. The accuracy of pilotage modeling of fast-time simulation is an issue that is separate from that of real-time simulation, and it is most often accomplished either by comparison with man-in-the-loop results or by simply presenting results to pilots familiar with the area in question and eliciting their comments and criticism. If the mathematical model of the ship track is known to be accurate, then the focus of the comparison is on the command sequence and timing produced by the autopilot. However, in the typical uses of fast-time simulation (that is, sensitivity studies to determine effects of changes in the waterway environment), a high degree of accuracy may not be needed.

VALIDATION

Because it is not possible to assess scientifically the accuracy of either the mathematical model (for fast-time or man-in-the-loop simulation) or the pilotage performance of individuals relative to the simulation, an overall validation of the simulation is typically conducted instead. Currently, a simulation is considered valid if pilots conclude that it accurately reproduces the modeled ship's behavior in a particular waterway (Eda et al., 1986; Hwang, 1985; Moraal, 1980; Puglisi et al., 1985b; U.S. Maritime Administration (MARAD), 1979; van de Beek, 1987; Williams et al., 1982). The idea is that if several pilots familiar with the waterway, the modeled ship, or both are all satisfied with the simulation, then there is reasonable confidence in the results. However, since this process is highly subjective, great care must be exercised to assure that preconceived views and experience with vessel behavior do not bias the evaluation of simulation results (J. P. Hooft, personal communication, 1992).

The process for developing a valid simulation (mathematical model plus pilotage model) is iterative. Various procedures are used to screen out unintended bias that might result from "ownership" in simulation runs, preconceived views about vessel behavior, and other performance and technical factors. Typically, simulator facilities use an interdisciplinary team approach for validation although the process is often not made formal in facility procedures. Pilots participating in preliminary simulations to validate simulator performance are either part of the validation team or provide information for the validation process. Other pilots, qualified facility staff or other experts are sometimes used to observe and evaluate simulation runs to guard against bias (J. P. Hooft, personal communication, 1992; Puglisi, 1985; Puglisi and D'Amico, 1985). The mathematical model of vessel behavior, the physical representation of the visual scene, and for fast-time simulation, the autopilot model, are modified until the validation team is satisfied that the simulator performs realistically. Modifications that are made represent an acceptable balance between theoretical considerations and practical experience. Considering the subjectivity that is involved in this process, the validation team must be carefully composed to include expertise for key governing factors in the simulation. Guidelines for validation team composition have not been established. The committee believes that for waterway design, expertise in vessel operations, hydrodynamics, mathematical modeling (of the physical environment and vessel behavior), and waterway design would be prudent.

As part of the iterative validation process, simulation facilities typically conduct interviews with pilots participating in real-time simulation before any simulations are performed and after each simulated transit. These interviews serve several purposes. The pilots' knowledge of subtle aspects of a waterway is especially important because these pilots may be the only source from which these features can be detected, as evident from simulation case studies (Appendix C). The interview agenda is structured to obtain the pilot's subjective interpretation of how the simulated vessel behaved under the conditions tested, such as realism of vessel response to wind. Their broad experience in shiphandling helps identify flaws in the mathematical model but, according to many simulation facilities, is less useful for identifying the cause of these flaws. Pilots also have individual general strategies for making a transit and use their critical visual cues to navigate according to these strategies. Omission of pilot-specific cues can lead to less than satisfactory simulations. Interviews help uncover omissions so that they can be corrected.

Finally, when analyzing a modification to an existing waterway, many facilities will first use a model of the existing waterway together with a model of a ship that currently uses the waterway. This simulates a situation familiar to the pilots and is useful for determining how much fidelity is needed to uncover errors in the basic mathematical model and to gain pilot confidence. This process also establishes a baseline from which to measure the effect of proposed changes. Introduction of either new ships or waterway configurations can then be made with greater confidence.

In developing a new harbor or waterway, rather than modifying or upgrading an existing one, it is more difficult to determine simulation validity, even subjectively. For new waterways, pilots (as well as others participating in the design process) have no local knowledge as a reference for assessing the simulator's performance. Similarly, pilots confronted with unfamiliar hull forms or vessel sizes would be constrained in their assessment of validity. When similar waterway configurations or ships are used elsewhere, experts familiar with them are sometimes invited to work with regional experts to determine simulation validity.

Considering the highly subjective nature of simulator validation in the maritime sector, can validation procedures be adapted from the aviation sector where simulation is widely used for design and training? The committee found substantial differences between simulation in the aviation and maritime sectors that make transfer of technology and procedures highly problematic. Aircraft flight simulators are an integral element of the aircraft design and evaluation process. They are used to support design decisions, assess the design validity, train pilots (initial and proficiency training), and support mission planning and analysis. For commercial aviation, both aircraft and flight simulators are certificated by the Federal Aviation Administration (FAA). Military operational flight trainers and weapon system trainers are accepted by the using organizations for new designs as the new aircraft enters flight operations. Attention to fine detail is integral to designing high fidelity simulation mathematical models and recreating the aircraft's cockpit. Very good simulation fidelity is obtained by using proper models of the aircraft's aerodynamics, propulsion system, control system, weight and inertias, and by including a high fidelity cockpit. The aircraft design process is structured to provide this information that concurrently supports development of flight simulators, including their validation (Appendix F). Pilots are brought into the airframe development process early to provide operational perspectives for airframe design, and participate in final testing of flight simulators, primarily adjustments to the mathematical model to gain pilot approval.

Unlike the aviation sector, shiphandling simulations are not developed as an integral part of vessel design. The design process for ships provides limited quantitative information for developing mathematical models for commercial ship behavior. Commercial ships are often one-of-a-kind or constructed in limited classes. Even if a class vessel, ship behavior varies significantly relative to other ships in the class with loading (which can radically change ship hydrodynamics) and other operational factors. Hydrodynamic testing is not performed extensively and aerodynamic testing is rarely conducted for commercial vessels. Ship trials data are available for certain vessels, but the operating envelope for testing is almost exclusively unrestricted deep water, providing no insight on variations (usually substantial) between ship behavior in deep and shallow water.

Substantial differences also exist between the aviation and maritime sectors relative to modeling the operating environment, particularly the effect of external boundaries. In particular, the forces on a ship are strongly affected by the details of waterway geometry. Modeling the aviation operating environment (such as the atmosphere and atmospheric disturbances) is more straightforward. Aircraft are mostly operated out of ground effect. Even if operated in ground effect for longer periods of time, modeling change in ground effect is much easier to predict than what is required to model the forces in relation to other vessels, shore structures, and bathymetry (which can vary dramatically in contour). Marine simulations also involve the effects of aerodynamic forces on a vessel that vary with draft and deck loads. Data on these effects are often insufficient or not available. Consequently, marine simulations for training and channel design lack the quantitative data that forms the basis for developing and validating aircraft flight simulators.

Characterizing ship-operating environment interactions remains a challenge in applying marine simulation technology to waterway design. Until a firm quantitative basis is developed, validation of marine simulations will continue to rely on subjective evaluations by expert marine pilots and other parties involved.

INTERPRETING THE RESULTS

As discussed in Chapter 3, the current state of practice in waterway design is to focus on ships and situations that will strain the safety of a waterway the most. Interpreting such results remains problematic. No formal basis was found to relate the results of simulation to a numerical measure of risk for the waterway. In other contexts, similar estimates are made for engineering projects by summing the products of probabilities of each possible accident in the project's lifetime and the cost of the consequence of the particular accident. This computation yields an estimate (expectation) of the risk exposure during the lifetime. Performing such a computation from the results of shiphandling simulation seems difficult, because only a few ships, pilots, waterway environments, and traffic situations are studied. The lack of objective validation of the simulation further compounds this analysis.

As a result, the current practice is limited to a subjective judgment that the waterway design is or is not satisfactory based on the limited simulations. Many facilities indicated to the committee that the judgments formed on this basis predict a much greater accident rate than is seen in practice. Whether this anomaly is due to mathematical model inaccuracy, lack of personal consequences to the pilot (including absence of liability and discipline that could result from mistakes in real life), or some other cause is not known. Often, simulation studies will be carried out on several alternative designs for the selected ship. Clearly, the judgments resulting from these comparisons may have more value, because fewer variables are introduced. That is, the trends of these comparisons may be more correct than their absolute tracks and may provide sufficient information for selecting one alternative over another.

SUMMARY

The accuracy of simulation for restricted shallow water almost certainly is lower than that obtained for deepwater maneuvering, because the latter situation has a much larger research base and does not require the use of so many heuristic models. Currently, no guidelines are available for assessing the accuracy required from the mathematical model to develop results useful for waterway design. Likewise, there is no numerical measure available for determining the accuracy of pilotage modeling. In particular, no guidelines are available for determining the level of simulation required for a particular situation or for an appropriate analysis of its results.

The state of practice is to use subjective measures (for example, interviews with pilots) to validate the overall simulation and subjective interpretations of the simulation results in terms of overall risk corresponding to the waterway design. Although questions about accuracy, validation, and interpretation cannot be resolved objectively, simulation has proven extremely useful in some applications (see Chapter 7).

7

Simulator Application in Harbor and Waterway Design

Shiphandling simulators have gained considerable acceptance world-wide as a useful aid in harbor and waterway design, albeit slowly and not yet universally. Detailed examination of representative simulation applications is useful to

- evaluate overall simulation usefulness,
- determine how simulation results were applied to the design process,
- develop an understanding of some of the problems encountered and how they were addressed,
- understand how simulation validity was verified, and
- assess the impact of different levels of simulator sophistication on simulation results.

Six simulation projects were selected for examination. These simulations were performed for projects at

- Oakland Harbor, California;
- Upper San Francisco Bay (Richmond Long Wharf), California;
- Grays Harbor, Washington;
- Norfolk-Hampton Roads, Virginia;
- Coatzacoalcos, Mexico; and
- Gaillard Cut of the Panama Canal.

The projects were selected for several reasons, including

• the variety of navigational-harbor design problems that the simulations addressed,
• the different levels of sophistication applied in several of the applications, and
• the firsthand knowledge of many of the simulations by the committee.

A summary of each of the six case studies is included in Appendix C. Each was reviewed with particular attention given to specific lessons learned that might be generalized for application to other simulations. These lessons, both technical and administrative, are included in the descriptions in Appendix C and are consolidated into the five findings that follow.

CASE STUDY RESULTS

Simulation results were used to reduce costs, increase ship safety, and reduce environmental risks.

In each of the six applications of simulators that were examined, the project sponsors were able to modify the waterway design and/or operation to achieve: significant cost savings, improvements in ship safety, and/or reduction in environmental risk. Cost savings were generally the result of reduced dredging or shifting of dredging activity to less costly sites. In one case, Coatzacoalcos, the cost savings resulted from being able to use larger ships safely without additional dredging.

Increased ship safety was achieved by identifying critical navigational areas during the simulation process. For example, in Oakland, significant safety benefits were derived by widening the bar channel and the entrance channel beyond the width initially proposed. Although this widening required an extra cost for additional dredging, it was offset by reduced dredging in other areas where the simulation had indicated it was not necessary. (Although the simulation was successful, port-sponsored project construction on an accelerated schedule has been discontinued because of legal constraints and the inability of the sponsor to develop a plan for disposal of dredged material that was acceptable to all parties [Appendix C]).

Environmental risks may be reduced by simulations in two ways. Improved ship safety contributes to a reduction in environmental risk because it reduces the probability of spillage of oil or other toxic substance that might result from ship groundings or collisions. Simulation can also reveal channel configurations that have smaller dredging requirements and which therefore minimize the environmental impact.

The committee cannot say with certainty that the cost, safety, and environmental benefits observed in these case studies would not have been achieved without the use of simulations, that is, if more intensive design reviews and audits had been conducted in the base case. However, it was observed that these benefits were not forthcoming before undertaking the simulations. Therefore, the committee finds it appropriate to credit the simulations for these benefits.

Simulation facilitated communication among the parties involved in a particular harbor and/or waterway project. These enhanced communications significantly affected the development of a successful, cost-effective design.

Successful harbor and waterway design involves the effective interaction of many different parties, including officials from federal and local governments, community and public interest (including environmental) groups, port operators, shipowners, hydraulic and civil engineers, naval architects and hydrodynamicists, environmental engineers, and pilots.

In the projects reviewed, simulation effectively focused the attention of these various parties on the harbor and waterway project. It provided a common framework for describing the project and the related problems as perceived by the various groups involved. In essence, it permitted these groups to communicate with each other with a common language and understanding that might not have been otherwise possible.

The use of simulation to focus communications demonstrated the potential of simulation to contribute measurably to successful, cost-effective harbor and waterway design and development. This benefit is separate from the research or engineering contributions that are usually expected of simulation.

Local pilots were regularly used by simulator facilities as the primary means of verifying simulation validity.

The issue of simulation validity received considerable attention throughout the study. During initial assessments, it was determined that an accurate mathematical modeling of the ship-channel interactive process was not possible, given current knowledge. The broad mathematical principles underlying the physical situation are generally well defined. However, a valid simulation in the strict engineering or scientific sense requires the measurement of environmental forces, their interaction, and the response of the ship to a degree that has never been attempted because of the great technical difficulty and costs involved.

In each of the six projects, local pilots were extensively used, not only as participants in the simulation process, but also, in essence, as the final arbiters of the validity of the simulation. Although the use of local pilots in this manner is not ideal because of the subjectivity involved, this limitation was recognized in the six projects and was addressed by incorporating greater safety margins than might have been necessary if the simulation validity could have been objectively measured. The risk exposure when applying simulation results, even with current knowledge, is considered less (at times significantly less) than that without simulation. Application of appropriate safety margins appears to be a reasonable way to deal with the current inability to measure objectively the validity of simulation.

Different levels of simulation were appropriate for different projects. Complex problems required sophisticated simulations. Sufficient information to resolve uncomplicated problems was obtained from low-level simulations.

Different levels of simulation were used in many of the case studies or, often, even within a single case study. Significant differences in the sophistication of the displays used for real-time simulation were noted between the various facilities. For example, some facilities had greater than 270° fields of vision, while others had only a straight-ahead display. Some facilities complemented the simulated radar screen with a bird's-eye presentation of the ship's position in the harbor, while others did not. Meaningful results were obtained from less-sophisticated simulations for certain problems. For example, fast-time simulations (no person in the loop) were used extensively for the Thimble Shoal Channel and Atlantic Ocean Channel simulations during the Norfolk-Hampton Roads project conducted at the Computer Aided Operations Research Facility (CAORF) for the State of Virginia and the U.S. Army Corps of Engineers. Navigation in these entrance channels does not require the use of any visual references from landmarks. Therefore, a real-time simulation with sophisticated displays was deemed unnecessary and costly. In other studies, high-quality visual displays were felt to be extremely important. This was the case for Grays Harbor, Washington, where it is necessary to navigate through two bridges that are offset from one another immediately after the vessel completes a sweeping turn in the river.

No comprehensive methodology was in place for assessing risk when interpreting and applying simulation results.

Simulation was demonstrated in application to be a source of guidance for the designer or user of a harbor and waterway. Simulation results must be interpreted and applied with care because their accuracy cannot be objectively verified. Although this limitation seemed to have been recognized by participants in each of the six projects, no comprehensive methodology was found to be in place for guiding designers on the establishment of safety margins or dealing with the uncertainties inherent in simulator results. This is a significant gap in the state of practice.

Simulations usually were not used in the early stages of the design process.

Simulations typically were used more often for design verification and refinement rather than for developing basic design parameters and limitations. This situation may change as harbor and waterway designers become more familiar with the usefulness of simulations and more skilled in their application. Some movement in this direction by designers seems to be taking place.

8

Research Needs

Shiphandling simulation is a high-level technology that is emerging as an important tool in waterway design. Previous chapters identified that many aspects of the present state of practice in the development, use, and interpretation of results for shiphandling simulations are, however, less than rigorous and scientific. As reflected in the case studies, the benefits of shiphandling simulation for visualization of waterway design problems and of consensus building are nonetheless great.

GAPS IN THE STATE OF PRACTICE

The committee's review of the use of shiphandling simulators for waterway design revealed an overall concern for validity and five specific technology areas that could benefit from substantial research. These areas, which are dependent on one another in an approximately sequential fashion, are the following:

- the level of accuracy required for the mathematical model,
- procedures for identifying and validating the mathematical model for ship behavior in restricted and unrestricted shallow water,
- information and procedures for determining the effect of fidelity of the pilot's visual and physical interface with the simulator on results of real-time simulations,

- a framework and standards for interpreting the results of simulation, and
- guidelines for the level and scope of simulation required in relation to the type of waterway design process.

A research program to fill these gaps is not presently being undertaken in the United States. Moreover, from a review of the state of practice of shiphandling simulation for waterway design in foreign countries, the committee found no evidence that such a comprehensive research program is being conducted abroad. Apparently, fundamental research on shiphandling simulation is rather moribund worldwide (for example, the privatizing of the Computer Aided Operations Research Facility (CAORF) at Kings Point, New York, has resulted in a shift in focus from fundamental research to contracted applied research and shiphandling training), although practical use of simulators for waterway design is growing. It is not clear whether or not congressional interest in research of marine simulation for operator training (generated by major tanker disasters) will result in a resurgence of basic operations research.

An original goal of this study was to develop guidelines regarding the appropriate level of simulation. Because substantial gaps remain in the five research areas necessary for developing such guidelines, the committee could not attain this goal.

Substantial improvement in knowledge and capabilities in each of the preceding areas holds promise for improving the confidence of practitioners and waterway designers in the results of simulations and, ultimately, for achieving the full potential of simulation. Although this study does not address use of shiphandling simulation for operator training, the basic questions concerning fidelity of simulation also apply where port-specific ship behavior is an element of the training regimen.

FUTURE RESEARCH

The committee has identified five specific areas for further research that would address the five technology areas defined above: 1) accuracy requirements for mathematical models, (2) identification and validation of the mathematical models, (3) effect of fidelity (visual and behavioral) on real-time simulation results, (4) interpretation of the results of simulation, and (5) guidelines for the required level and scope of simulation.

Fidelity Requirements for Mathematical Models

As discussed in Chapter 6, shiphandling involves intelligent feedback to available cues (either in real time using a human pilot or in fast time using a sophisticated pilot model). In either case, the pilot or autopilot corrects for errors, whether they are due to real effects or errors in the mathematical model. This situation appears to reduce the demands for accuracy on the mathematical model for ship behavior (over that required for an open-loop, dead-reckoning model), but there is little or no information in the literature to document this conjecture or to indicate what level of model accuracy is required. Research could be conducted to determine the sensitivity of the results of simulation for waterway design to either the framework for the mathematical model or the accuracy of the coefficients used in conjunction with this framework.

The mathematical model frameworks in typical use differ little, if at all, in their linear terms. The differences, where they occur, exist in the number and arrangement of higher order terms. A better understanding is needed about the requirement for accuracy of the various coefficients in typical mathematical frameworks, in particular, the coefficients that characterize the effects of small under-keel clearances and the interactions with varying banks and currents. Identification of these coefficients is exceedingly difficult and therefore expensive.

Sensitivity studies constitute a necessary preliminary for the remainder of research opportunities, but need not represent a significant investment. Fast-time simulation is ideal for this purpose because it is repeatable and does not include the variability inherent in human pilots. Examples exist of complete mathematical models for ships in restricted waterways. Investigations of the adequacy of the framework can be based on the use of different known models for one ship type and models for several representative waterways. Investigations of the accuracy of the coefficients probably will involve systematic perturbations of the mathematical models for several different ship types and for several representative waterways.

Identification and Validation of the Mathematical Models

If, presumably, the above-mentioned research determined the level of accuracy needed in coefficients of a mathematical model, the problem would remain of identifying these coefficients. Chapter 5 revealed that considerable weaknesses exist in the identification of hydrodynamic coefficients for use in a mathematical model. Although scientific means are available for performing accurate identification, the expense would be prohibitive for tests to characterize the behavior of just one ship in all possible small under-keel clearance and bank situations. New, less costly approaches are needed to overcome this gap.

Developing this information by computational fluid dynamics (CFD) is beyond present capabilities, although the use of CFD methods is rapidly expanding. Physical modeling techniques and the scaling relations necessary for their performance exist and have been known for some time. However, physical modeling is limited in three ways. First, the number of tests necessary to characterize the hydrodynamic forces on a single ship in a restricted waterway with a shallow bottom and banks is very large, and as a result, the costs would be large. Second, only a few facilities in the world have flat-enough bottoms to perform model tests in shallow water comparable to realistic under-keel clearances. None of these facilities are in the United States. Finally, it is impossible to scale viscous effects in small-scale model tests, and there is reason to believe that this factor is important in the case of small under-keel clearances.

Even if extensive model tests were performed, validating the resultant mathematical model would be even more challenging. To date, validation by comparison with full-scale measurements of ship trajectories in restricted waterways has been limited to only a few cases. Even so, the most extensive and most scientific of these (the *Esso Osaka* trials) did not involve small under-keel clearances comparable to typical waterway situations or the influence of banks.

Validation methods of deepwater maneuvering predictions that are based on full-scale maneuvering trials have often been incorporated in the delivery trials of new ships. However, these trials usually have been aimed at simple turning performance and steering stability. Similar trials in shallow or restricted waters common in waterway design have not been performed for reasons of safety. This constitutes a significant gap in the validation tools for waterway design. Associated research opportunities include

- the development of new, efficient techniques that could reasonably be expected to identify numerical coefficients for a mathematical maneuvering model for restricted waterways to the level of accuracy required, and
- the development of techniques for safely conducting full-scale tests in typical waterway situations and for analyzing the results to calibrate and validate the mathematical models.

The committee anticipates that these techniques would be tailored specifically to the needs of waterway design. In particular, it is anticipated that the required accuracy of the coefficients may be less than that achievable by classical model tests, and that new, innovative, and more economical techniques would result from exploiting this requirement. The committee further anticipates that the identification and validation would likely be a combination of theoretical results and model tests (perhaps involving systems identification methods or CFD approaches).

Effect of Visual and Behavioral Fidelity on
Real-Time Simulation Results

Conventional wisdom states that the higher the visual fidelity of the simulator, the more useful will be the results of real-time simulation. Although some studies have addressed individual aspects of fidelity, proof of this conjecture does not exist. The importance of developing more information concerning visual fidelity is driven by two considerations. First, with the cost of computation now decreasing dramatically, visual fidelity is no longer the principal determinant of the cost of a simulator facility. Second, new shipboard instrumentation is developing at such a pace that some existing studies of fidelity may no longer be relevant.

Bird's-eye view displays have often been the only displays available in low-fidelity simulators, a feature that is considered by some to be a defect because the displays provide the pilot with more information than would be available on a real bridge. The development of differential GPS (global positioning system), digital chart data, and inexpensive on-board computer graphics equipment (relative to the cost of the ship and cargo) have made accurate bird's-eye view displays a reality. Electronic chart systems can include all the aids to navigation and other waterway information. In the future, integrated bridges, some with piloting expert systems (that is, artificial intelligence decision aids), together with normal shipboard sensors may produce other displays that communicate real-time decision-making information to the pilot. Integrated bridges are presently available on only a small number of vessels worldwide. An associated research opportunity is to determine the presence and fidelity of such systems on real-time simulations used for waterway design. The effort would investigate the potential role and efficacy of new instrumentation available to pilots and the extent to which this needs to be represented in marine simulations.

Interpretation of Simulation Results

Chapter 3 stated that shiphandling simulation is based on the assumption that a small sample of simulations using one or two ships, a few environmental conditions, and a few pilots will provide meaningful information for use in waterway design. Because the accuracy of the mathematical models of ship behavior is in question, the setup of a simulator facility is so costly, and the collateral benefits of simulation (for example, consensus building) are often an important objective, less emphasis has been placed on developing a formal framework for interpreting marine simulation results than on simulation in other industries.

Elaborate frameworks (usually statistical measures) have been developed for quantifying the performance of many engineering simulations in other disciplines and for use in design. Typical among these are simulations of the behavior of telephone networks, traffic congestion on highways, and flow of products in oil refineries. In these fields, the characteristics of the system elements are well known, but the system is subjected to random demands that stress it in complicated ways. Statistical measures are used to relate the design parameters (which reflect construction costs of the system) to risk of failure and its consequences (the contingent costs of the system).

Other types of simulations use quite different frameworks for analyzing performance. For instance, simulators designed to train personnel in the art of aircraft handling include both objective measures, for instance calibration against measured performance characteristics (such as turning circles) and subjective measures, such as pilot confidence, that the simulator behaves like a real airplane (Appendix F).

Unfortunately, the use of shiphandling simulators for waterway design does not fit comfortably in any of these molds. Shiphandling simulation is in many ways more difficult and complicated than the examples cited. As with aircraft simulators, marine pilots must feel confident that the "feel" of the simulated ship is like a real ship. However, information to calibrate the model, such as performance in very shallow water or near banks, is not known with scientific accuracy for any commercial ship. As with road traffic simulations, the quality of pilots and the number of different ship types and their performance vary greatly. However, an equivalent to the considerable data that characterize vehicle behavior on a highway does not exist for ships. The highway problem is also simpler in another way: there is no analog for the changes in steering performance needed in ships due to changes in under-keel clearance or banks. The basic problem in shiphandling simulation arises because the sample size in the simulation is so small restricting the application of classical inferential statistics.

Needs for future research include the following:

• The development of a framework to interpret results of a small sampling of simulator runs in terms of the quantities that affect waterway design. This framework could include, for example, a numerical estimate of the significance of results and confidence bounds on the predictions of swept paths measured in the simulation.

• The development of a procedure for estimating risk of accident associated in a particular waterway design and for estimating the consequences resulting from potential accidents.

Guidelines for the Required Level and Scope of Simulation

A synthesis of the results of research suggested above could yield considerable insight into the level and scope of simulation required for the waterway designer. However, simulation is a sophisticated technical discipline that relates to only one aspect of waterway design, and it is outside the typical focus and training of the waterway designer. The planning representatives of local project sponsors who must approve the expense of simulation are even further removed from this expertise. Therefore, an associated research opportunity exists for synthesizing information from the previously mentioned research and transforming it into a set of guidelines that could be used by the waterway designer and the sponsor to select appropriate simulation studies for a particular waterway design.

The committee found that such guidelines do not exist. As a result, waterway designers and their sponsors have little basis for selecting one simulator over another and for selecting the scope of simulation studies to be performed. Further, a set of guidelines based on a firm scientific footing could permit more rational decisions regarding when and to what extent simulations should be performed for given waterway projects.

STRATEGIES FOR IMPLEMENTING A RESEARCH PROGRAM

Support for basic research on shiphandling simulation has withered within the past decade. Only the U.S. Army Corps of Engineers (USACE) has a modest, project-oriented program in this area. As a consequence, the number of persons within the interested federal agencies experienced with shiphandling simulations has declined. However, the services of a substantial number of ship hydrodynamicists, both internal and external to the federal government, might be applied to fundamental research. Several experimental facilities exist worldwide that could be used to conduct elements of the research. (A facility catalog of ship hydrodynamic facilities is maintained by the International Towing Tank Conference.) One notable limitation is the inability of existing facilities to scientifically validate the scaling of mud behavior from model scale to full-scale (that is, reproducing on a model scale a material that would emulate behavior of bottom sediment) when testing ship maneuverability in situations with very small under-keel clearances.

The research program necessary to put shiphandling simulation for waterway design on a firmer scientific basis, thereby greatly increasing the confidence of the entire maritime community in the usefulness of the technique, would be ambitious, expensive, and long range. The committee believes such a program would require about 10 years of dedicated effort and about $15-30 million in research funds. However, costs in this range are modest relative to the annual investment in port facility capital improvements and in waterway construction, operations, and maintenance. Developing a strategy for the research program would entail addressing sponsorship as well as the resources necessary to conduct the research, including skills and facilities.

The USACE is the government agency charged with primary responsibility for waterways development, including waterway design, permitting, dredging, and disposal of dredged materials. With this leadership role comes implied responsibility for organizing and coordinating research needed for waterways development in the United States. USACE operates a modest computer-based simulator, has a small pool of technical expertise, and has used these resources in about 40 waterway design studies. With regard to basic simulation research, a technically broad scope of effort would be required. USACE, by practice, principally conducts and has good experience with limited-term, project-oriented research for civil works. Present USACE technical resources do not appear sufficient to undertake or guide the multi-year research program that is needed to improve the scientific basis of simulation technology. In the committee's opinion, USACE would need to augment its technical base with experts from industry, especially in the areas of ship hydrodynamics simulation-based research, and human factors.

Other entities that have interest in waterway design (and in some cases operator training) and are potential beneficiaries of improvements in the national simulation capability include the U.S. Coast Guard (USCG) and U.S. Maritime Administration (MARAD) (both of which have previously been involved in simulation research), project sponsors, shipping companies, and operators of port facilities. Resources of these entities vary widely. The committee found that none of the non-federal entities appeared to have either sufficient focus or in-house capability to independently direct even a small part of the research program.

Two implementation strategies appear feasible. The federal government could fund the research in support of a national interest in maintaining competitiveness in international commerce. USACE could undertake, plan, and coordinate a government research program, which includes participation (and perhaps cost sharing) by other involved government agencies such as MARAD, the USCG, and perhaps the U.S. Navy (USN), in a supporting role.

An alternative would be to establish a government-industry research consortium that included key components of the U.S. maritime industry in both sponsorship and technical support capacities. Nongovernmental participants could include waterway designers, port operators, shipowners and operators, and pilot associations. This approach would have the advantage of involving all the direct beneficiaries to support the research, although coordination of such a body could prove cumbersome.

This research program could take full advantage of available expertise and capabilities at existing simulation and research facilities throughout the United States to ensure that the selected research plan is focused on the areas of greatest need, is sufficiently comprehensive, and is cost-effective.

In an environment in which available funding is likely to remain limited, it is essential to ensure the maximum cost-benefit ratio of all research conducted.

SUMMARY

The development of meaningful guidance for waterway designers and sponsors on the use and applicability of shiphandling simulators for waterway design is inhibited by gaps in knowledge and capabilities in several critical areas. Because of the complex scientific basis for simulation and the hardware associated with it, the research required to fill those gaps is essential for full utility and acceptability of the technique, albeit its expense. Currently, no government agency, commercial enterprise, or research organization has undertaken or appears ready to undertake a dedicated research program on shiphandling simulation. Such a program could be a joint government-industry initiative, perhaps dovetailed with research that may be needed to establish shiphandling simulation as a fully accredited shiphandling training aid. USACE would be an appropriate organization to coordinate needed research because of that agency's primary role in waterway development.

9

Conclusions and Recommendations

The design of waterways affects the nation's economy, the safety of ships and their crews, the inhabitants near waterways, and the natural environment of waterways. Over the past two decades, the use of shiphandling simulation to achieve refinements in waterway design not verifiable with other design tools has significantly increased. However, use of simulation in this way has been incorporated in only a small portion of the total number of waterway projects.

DOES SIMULATION WORK?

Shiphandling simulation has been used effectively as a design tool by planners and engineers to aid substantially in waterway design. The committee found the following:

- Simulation can be and is used during early and later stages of the design process to answer critical design questions, including those raised during permitting. Early use of simulation is especially important in cases where it can be used on a recurring basis throughout the design process.
- Pilot acceptance of simulations during validation and study trial phases indicates reasonable success in re-creating a realistic piloting experience.
- Simulation offers a systematic means for capturing the complexity

84

of a waterway layout, the physical environment, and operational factors of a waterway design in an integrated and visible fashion.

• Simulation enhances communication between design participants. It brings together various constituencies with interests in waterway design, thereby providing a unique, common forum and framework for discussion and decision making.

WHEN SHOULD SIMULATION BE USED?

RECOMMENDATION: *Practitioners should use simulations in all waterway design problems where ship operational risk is important. Furthermore, it is advisable to use simulation where optimization is an objective.*

Although cost, significant gaps in knowledge and capabilities, and lack of confidence inhibit wider use of simulation, the efficacy of applying ship-handling simulators as a design aid has been proven in practice. In spite of all the uncertainties that exist in terms of modeling and interpreting simulation results, the demonstrated benefits of simulation for a wide variety of projects more than adequately justifies its use as a standard practice in waterway design.

Simulation should be used in the following situations:

• When vessel operational risk is a significant design issue. Representation of human pilot skills and reactions in the prediction of vessel behavior in a proposed waterway is unique to shiphandling simulation. Differences in risk under various critical environmental conditions can be identified. Requirements for aids to navigation to further reduce risk can also be assessed.

• When cost and design optimization is an issue. The effect on risk resulting from variations in many design factors that define a waterway can be evaluated. This capability is important for assessing the components of life cycle costs. Simulation is particularly useful for assessing operational differences between design alternatives.

• When competing interests among technical and nontechnical participants in the waterway design process are an issue. Simulation provides a unique way to bring critical and contentious aspects of the design into sharp focus. The consequences of what participating parties are interested in can be acquired and displayed in formats that do not require technical expertise to assimilate and understand.

Because elements of these three issues are frequently associated with most waterway designs, shiphandling simulation should be developed as a standard tool for use in the waterway design process. The level of sophistication of simulations needed for this process depends on the particular design. However, guidelines for what level is appropriate for a given situation are not available within the current state of practice.

HOW CAN SIMULATION BE ENHANCED AS A DESIGN AID?

RECOMMENDATION: *Simulation facility operators should establish a formal validation process that uses a carefully composed, interdisciplinary validation team to assure that key governing factors are adequately addressed and to provide consistency in the validation process.*

Simulation is not used more often by designers for three principal reasons: costs and schedule of simulation, lack of confidence in the results, and lack of awareness of simulation as a design tool.

Costs of conducting simulation studies presently inhibit the use of simulators in the design effort. The cost of the simulator itself, because of advances in computer technology, is no longer the limitation it was just a few years ago.

The state of practice of shiphandling simulation for waterway design varies widely. No agreement exists among practitioners on the minimum requirements for simulator fidelity for a given application. From examination of previous applications to waterway design, it is evident that a significant level of confidence in the application of shiphandling simulation to waterway design is not uniformly shared by all waterway design participants.

This lack of confidence revolves about questions of overall fidelity and validation. The components where fidelity is questioned are mathematical models of ship dynamics, waterway data bases, and visual displays. The behavior of ships with small under-keel clearances is especially not well understood nor well represented in existing models. Increasing the level of user confidence and acceptance will require development and validation of more robust mathematical models. Other factors that inhibit simulation include:

- the lack of a formal, objective method to validate the model and
- the lack of an accepted scientific framework for interpreting simulator results for waterway design. No consistent means exist for extrapolating results from the small sample of real-time runs to a prediction of the performance of the design over the life of the waterway.

To make simulation a more attractive design option, basic research should be conducted to resolve confidence issues and provide the capability for more effective simulation. A single, cohesive research program, focused on identified research needs, should be defined and managed as a coordinated effort that draws on the best technical expertise available within the waterway design and simulation community. Multi-disciplinary involvement in improving simulation capabilities would help increase confidence by the port and maritime transportation communities in simulation as a design and evaluation tool for waterways. Multi-disciplinary participation can be improved immediately by establishing formal validation processes that include essential operational and technical expertise in carefully composed interdisciplinary validation teams.

ESTABLISHING A RESEARCH PROGRAM

RECOMMENDATION: *A systematic program of research designed to put simulation on a firmer scientific footing and to develop means for guiding its use and interpretation should be undertaken as a joint government-industry initiative. It should be coordinated by the U.S. Army Corps of Engineers and should include participation by pertinent federal agencies and the port and marine transportation communities. The research needs identified in Chapter 8 should be the central elements of the research program designed to*

- *assess the need for fidelity in mathematical models and simulator hardware,*
- *develop accurate means to identify the elements in the mathematical model, and*
- *develop means to interpret the results of simulation.*

The research program should improve the design tools needed to develop safe and cost-effective waterways. The program would be expensive and would require long-term funding. The size and scope of the research program is beyond the budget allocations from government agencies with responsibilities for waterway design and operation. Such a program would also have a cost and time frame that would be beyond incremental improvements of current programs. Implementation will require recognition of these research needs by Congress and the Departments of the Army and Transportation as a national priority to assure competitiveness with national research needs in other fields.

The U.S. Army Corps of Engineers, in view of its designated responsibility for waterway design, should take the lead in coordinating the research program. Such a research program should be carried out on a cooperative basis by all interested parties and beneficiaries. Program participants should include the U.S. Coast Guard, the U.S. Maritime Administration, and other organizations within the marine transportation and port communities. Funding support should be provided by the federal government because of national interests in ports and waterways and by beneficiaries in the port and marine transportation communities. Development and execution of the research program should take advantage of available expertise and capabilities of the existing research and simulation facilities across the country.

Appendixes

A

Committee Member Biographies

WILLIAM C. WEBSTER, *Chairman,* is full professor of naval architecture and associate dean of engineering for student affairs at the University of California at Berkeley. He is a leading national authority on ship maneuvering and has been heavily involved in the hydrodynamic aspects of ship model testing and in the mathematical developments underlying ship-handling computer simulation. Dr. Webster has published on both the hydrodynamic and operational aspects of ship maneuvering, has served as member and chairman of the Marine Board, and currently is a member of the Commission on Engineering and Technical Systems. He received his B.S. degree in naval architecture and marine engineering from Webb Institute of Naval Architecture and his M.S. and Ph.D. degrees in engineering science/naval architecture from the University of California at Berkeley.

WILLIAM A. ARATA is a state and federally licensed pilot with Biscayne Bay Pilots in the Port of Miami, Florida. Prior experience includes 20 years in the U.S. Navy where he served primarily in the submarine fleet in capacities from watch officer to commanding officer. He also directed naval analysis programs of the Office of Naval Research and submarine electronic programs for the deputy chief of Naval Operations, Submarines. Captain Arata received a B.S. degree from the U.S. Naval Academy, a B.S. degree in mechanical engineering from the U.S. Naval Postgraduate School, and a M.S. degree in business administration from the George Washington University.

RODERICK A. BARR is a principal in Hydronautics Research, Inc. He has 30 years experience in marine hydrodynamic research and development, analysis, and design. Previously, he was a principal in Tracor Hydronautics. His experience includes development of theoretical methods; computer-based, time-domain simulation studies; model testing; and design studies. Dr. Barr has lectured on ship hydrodynamics at the Catholic University, and authored numerous technical papers and reports. He received a B.S. degree in naval architecture and marine engineering from Webb Institute of Naval Architecture, a M.S. degree in mechanical engineering from the University of Maryland, and a Ph.D. degree in naval architecture from the University of California at Berkeley.

PAUL CHILCOTE is senior director for planning, budget, and environmental affairs at the Port of Tacoma. He is responsible for coordinating port development into long-range planning, management of the port's Capital Improvement Program, and environmental management. Previously, Mr. Chilcote was a consultant for the State of Washington, specializing in port, rail, and marine transportation; manager, Intermodal Market Planning-International with the Southern Pacific Transportation Company; and senior trade analyst and senior long-range planner with the Port of Seattle. Mr. Chilcote received a B.S. degree in international/urban geography and history from Long Beach State University and a M.S. degree in economics/transportation geography from Oregon State University.

MICHAEL DENNY is president of ShipSim Corporation. During the study, he was systems architect with the Data Systems Division, Grumman Corporation, where he managed the technical development program, concentrating on systems architecture, expert systems, and artificial intelligence. Earlier, Dr. Denny was program manager with Ship Analytics, Inc., where he managed port and waterway design, vessel navigation bridge procedures, and human performance experiments at the Computer Aided Operations Research Facility, Kings Point, New York. His work has spanned both shiphandling simulation for waterway studies and leading edge computer simulation developments with human interaction. Dr. Denny has also served as assistant professor in the psychology department at Michigan State University. Dr. Denny received his B.S., M.A., and Ph.D, degrees in experimental psychology from Michigan State University.

FRANCIS X. NICASTRO is coordinator, industry affairs, with the Transportation Department, Exxon Company International. Previously as manager, Chartered Ship Operations and Port Services, he was responsible for the Exxon-sponsored simulation study of the navigability of tankers in the oil port of Coatzacoalcos, Mexico, which was conducted at the Computer Aided Operations Research Facility. Other technical and managerial service with Exxon Company International have included chartering manager, commercial support manager, and vice president and manager, New Construction Office, Kobe, Japan. Mr. Nicastro has served on various technical panels of the Society of Naval Architects and Marine Engineers. He received a B.S. degree in naval architecture from Webb Institute of Naval Architecture.

NILS H. NORRBIN is internationally known for his work in ship hydrodynamics, maneuvering, and shiphandling simulation. Dr. Norrbin retired from the Swedish State Shipbuilding Experimental Tank (SSPA) in 1991 as senior scientific advisor and project manager. Previous positions with SSPA included head, Dynamics Division, and head, Ship Dynamics Division, of the Research Department. Dr. Norrbin is a fellow of the Royal Institution of Naval Architects and a member of Fachmitglied, Schiftbautechnische Gesellschaft. He has served on many international committees concerned with ship maneuvering, has pioneered research in the effects of realistic ship channel boundary effects on navigation, and was visiting professor of naval architecture, Department of Ocean Engineering, Massachusetts Institute of Technology. Earlier work included naval design with the Royal Swedish Naval Administration. Dr. Norrbin received a M.Sc. degree in naval architecture from the Chalmers University of Technology, a Technical License in Ship Hydraulics and Mathematics, and the Dr. Technology degree from the Royal Institute of Technology, Stockholm.

JOSEPH J. PUGLISI is associate director, Office of Computer Resource, the U.S. Merchant Marine Academy. He is responsible for all academic and administrative computing at the academy including integration of computers into the curriculum. Previously, Mr. Puglisi was managing director, Computer Aided Operations Research Facility (CAORF), where he was responsible for over 100 shiphandling simulation studies leading to formal reports. He directed the development of upgrade plans for CAORF, including marketing, research, system expansion, design and engineering, manpower, and funding. He subsequently coordinated field-level components of the privatization of CAORF. Mr. Puglisi received a B.S. degree in electrical engineering from the City College, City University of New York, and a M.S. degree in electrical engineering from New York University. He is pursuing advanced studies at the University of Wales.

LEONARD E. VAN HOUTEN is a consulting engineer with over 35 years experience worldwide in charge of planning, design, and construction of industrial facilities for marine transportation, oil and gas production, mining, heavy manufacturing, and defense-related activities. He has led development of waterway improvement projects in 45 countries representing all continents and a complete range of climatological conditions. Mr. Van Houten previously was a member of the Marine Board's Committee on Sedimentation Control to Reduce Maintenance Dredging of Navigation Facilities in Estuaries and is active in a number of technical and professional societies. He received B.S. and M.S. degrees in civil engineering from Rensselaer Polytechnic Institute.

JAMES A. VINCENT manages the Aeronautical and Marine Systems Department with Systems Control Technology. His experience is in flight mechanics, with specialization in system identification air vehicle mathematical modeling, control system design, simulation testing, handling qualities analysis, and wind tunnel testing and aerodynamic configuration development. Previously, Mr. Vincent served with the Boeing Company where he validated flight simulation by actual flight tests. He received B.S. and M.S. degrees in aeronautical engineering from the University of Colorado.

B

Design Elements of
Waterway Development

The waterway development process in the United States follows procedures prescribed by the U.S. Army Corps of Engineers (USACE). Six phases are organized in a logical progression:

- reconnaissance,
- feasibility,
- preconstruction engineering and design,
- real estate acquisition,
- construction, and
- operation and maintenance.

Although a progression is indicated, considerable overlap occurs at various stages, particularly if a project or its design is challenged after the reconnaissance phase is completed.

Prior to the reconnaissance phase, local interests determine whether a project is needed, what the project should entail, and for which elements interested parties are willing or able to become project sponsors. The usual procedure is for the local sponsor to petition the U.S. Congress and USACE for authority and funds to conduct a study. If the petition is successful, studies constituting the reconnaissance phase are conducted by USACE at full government expense. The objectives are to define the opportunity for the project; assess support; determine apparent costs, benefits, environmental impacts, and potential solutions; and estimate cost and time for the next phase. USACE usually brings others into the process, but not necessarily all parties that subsequently may be determined to have an interest.

The project may or may not proceed into the feasibility phase. Feasibility phase costs are shared equally by the sponsor or sponsors and the federal government. In this phase, alternate plans are identified and evaluated, leading to a full description of the project. All aspects of the project are supposed to be examined and all potential participants brought into the process. The desired product from this phase is a final project form, based on consensus insofar as practical, that is acceptable to all interested parties.

If the logical progression were followed exactly, a design would be fixed by the end of the feasibility phase after consideration and analysis of all reasonable alternatives, benefits, and impacts (Olson et al., 1986). This design would then be used as the basis for necessary permits and other following elements in the process.

In practice, the selected design may be challenged, for technical, social, or environmental reasons up to and including the construction phase, to address unrecognized flaws or further address competing objectives. Resolution of challenges may result in accommodations affecting the technical integrity of the original design solution, which necessitates further studies, data collection, and design work. Delays in completing the process may affect the availability of options selected and resources available (see Kagan, 1990).

C

Practical Application of Shiphandling Simulators to Waterway Design

CASE STUDY
COATZACOALCOS, MEXICO
COMPUTER AIDED OPERATIONS RESEARCH
FACILITY, 1980-1981

PROJECT DESCRIPTION

Exxon Corporation sponsored a simulation study at the Computer Aided Operations Research Facility (CAORF) to determine the maximum size oil tanker that could safely load at Coatzacoalcos, Mexico. The exit from the loading docks requires passage through a narrow, 330 foot wide channel cutting obliquely across the Coatzacoalcos River. Very high river currents were reported during the rainy season. The harbor chart is shown in Figure C-1.

SIMULATION DESCRIPTION

CAORF, a government-owned but privately operated facility located on the grounds of the U.S. Merchant Marine Academy, Kings Point, New York, was selected to conduct the simulation study. Real-time simulations were conducted using local pilots. Simulations compared the performance of the larger ships proposed for this service with that of ships actually using the

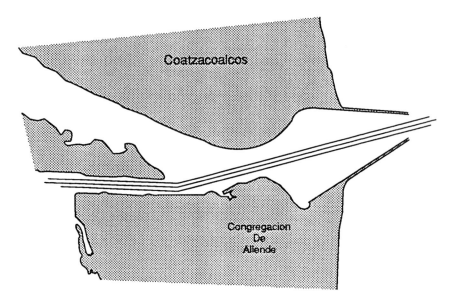

FIGURE C-1   Coatzacoalcos Harbor.

port. Generalized mathematical models were used in the simulations based on material available for similar ships.

Field visits were made to develop visual effects for the simulation, to observe the performance of ships using the port, and to measure environmental effects (river currents). Measured river current velocities were found to be significantly less than those reported by local pilots and harbor authorities. As a result, a considerable amount of additional testing and interpolation of results was necessary.

SIMULATION RESULTS

The simulation indicated that tankers about 10 percent larger (70,000 deadweight tons {DWT}) than those currently used could be safely loaded at this port. The larger tankers required a slightly greater under-keel clearance than the smaller tankers during the rainy season (U.S. Maritime Administration, 1980, 1981).

PROJECT IMPLEMENTATION

Larger tankers regularly began using this port in late 1981. The cost of simulation was recovered in less than 2 months of operation of the larger ships.

LESSONS LEARNED

Design Team

A shiphandling simulation is not a simple engineering project. Success depends on assembling a team of qualified and imaginative professionals (hydrologists, hydrodynamicists, pilots, simulator operators, and end users) and managing that team in a way that encourages each member's full contribution.

Pilot Participation

Local pilot participation was important to verify qualitatively the accuracy of the simulation. In this study, changes in the treatment of bank effects were made as a result of pilot advice.

Operating Parameters and Costs

Significant economic savings may be gained from relatively modest changes in permissible operating parameters.

Data Validation

Assessments of environmental data that are not supported by accurate field measurements should be carefully weighed before they are accepted.

CASE STUDY
NORFOLK/HAMPTON ROADS, VIRGINIA
COMPUTER AIDED OPERATIONS RESEARCH
FACILITY, 1980-1986

PROJECT DESCRIPTION

The State of Virginia sponsored a simulation study to improve existing channel designs so as to permit deep-draft coal colliers of 225,000 DWT and 55-foot draft. The objective of the project was to make Hampton Roads ports (Figure C-2) more competitive in the world coal market.

SIMULATION DESCRIPTION

This project was the first large-scale, real-time and fast-time simulation undertaken by the U.S. Army Corps of Engineers (USACE) for a U.S. port.

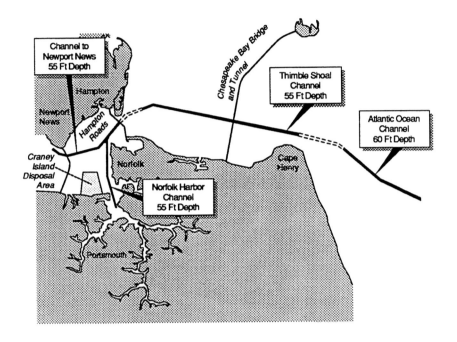

FIGURE C-2   Hampton Roads (*USACE*).

Simulation costs were about $1.3 million. The simulation was performed at the Computer Aided Operations Research Facility (CAORF). Thirty state-licensed pilots from the Virginia Pilots Association and 19 docking masters affiliated with tugboat companies participated in the real-time simulation for the Norfolk and Hampton channels. Fast-time simulation was used to evaluate ship maneuverability in Thimble Shoal Channel and the Atlantic Ocean Channel. The Chesapeake Bay physical model as well as numerical models developed by the USACE Waterways Experiment Station, Vicksburg, Mississippi, were used to evaluate the environmental conditions in these channels (USACE, 1986a).

SIMULATION RESULTS

The simulation indicated that the new channel design would permit bigger and deeper colliers to sail safely from Hampton Roads ports. The simulation recommended maintaining some channel widths as they were initially designed and reducing others, for example, from 1000 feet to 650 feet. Savings of over $100 million were projected as a result of reducing the amount of required dredging (U.S. Maritime Administration, 1985).

PROJECT IMPLEMENTATION

The project is to be implemented in two phases: channel widening and depth to 50 feet, then depth to 55 feet if economically required. Phase 1 was completed in 1988; Phase 2 is under review.

LESSONS LEARNED

Consensus Building

As a result of the simulation, the channel design was extensively coordinated with the U.S. Coast Guard, U.S. Maritime Administration, Virginia Pilots Association, Virginia Port Authority, USACE, and others. This cooperation resulted in an effective channel design effort with significant cost savings.

Effectiveness as Design Aid

The simulation demonstrated that an asymmetrical channel design would not impair the safe movement of large, deep-draft vessels in confined channels. The combination of real-time and fast-time simulation provided an effective tool for analyzing various channel design options.

Risk Assessment

Simulation runs were not based on worst-case conditions. As a result, vessels may be at risk when encountering these severe conditions, and operating practices need to be developed by the port users to account for these untested risks.

CASE STUDY
JOHN F. BALDWIN, PHASE 2
(RICHMOND LONG WHARF)
WATERWAY EXPERIMENT STATION, 1983-1984

PROJECT DESCRIPTION

The simulation study was designed to verify the validity of the final design for a major ship channel improvement project in Richmond, California, in the Upper San Francisco Bay (Figure C-3). A design had been completed that involved deepening the Southampton Shoal Channel (connecting channel) and maneuvering areas to the Richmond Long Wharf in order to permit the discharge of fully loaded 85,000 DWT tankers and partially loaded 150,000 DWT tankers at this port. The design's suitability for large containership operation into the Richmond Harbor navigation channel was also to be checked.

SIMULATION DESCRIPTION

This simulation was undertaken by the USACE Waterways Experiment Station (WES), located in Vicksburg, Mississippi. It was one of the earliest simulations conducted at this facility. Only modest graphic display capability was available.

Generalized mathematical models were used in the simulation based on material available for similar ships. Environmental data were either measured from scale hydraulic physical models or based on local records as appropriate. Several simplifying assumptions were made in setting up the simulation in the interest of saving time and reducing costs.

Real-time simulations were done for both tankers and containerships. Only limited use was made of experienced pilots, with 36 of the 41 simulation runs being done by WES engineers (Converse et al., 1987; Huval et al., 1985).

SIMULATION RESULTS

The simulation indicated that the initial design was suitable for use by the intended tankers and the smaller (638 foot length overall [LOA]) containerships. However, operation of the larger (810 foot LOA) containerships was not recommended with this design because of tidal currents expected in this port. Significantly less dredging was found to be necessary than was initially proposed for the maneuvering area.

The simulation, which cost only $110 thousand, resulted in a $1.8 million direct savings in dredging costs out of a total project cost initially estimated to be $12.8 million. A value engineering analysis, performed later using additional simulation testing, produced an additional $2.2 million in dredging savings.

PROJECT IMPLEMENTATION

Southampton Shoal Channel and the Maneuvering Area were dredged in 1986 in accordance with the USACE design as revised by the simulation results and the value engineering analysis. Satisfactory operation is reported for the intended tanker traffic and smaller containerships. Additional studies for improvements to Richmond Inner Harbor and approaches are under way.

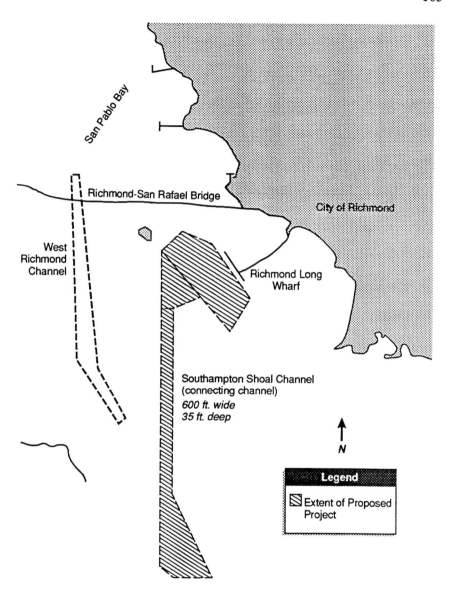

FIGURE C-3   Study area for John F. Baldwin (Richmond Long Wharf), Phase II (*USACE*).

LESSONS LEARNED

Design Team Skills

This simulation was successfully used for design verification and cost reduction despite its relatively modest level of sophistication and the limited simulator operating experience of the WES staff for this early simulation. These limitations were addressed by using appropriate safety reserves when applying the simulation results.

The success of this particular simulation owed much to the skill of the simulation-design team in simplifying the simulation so as to obtain meaningful results in a timely manner. Their success clearly illustrates the importance of having a skilled team doing simulations.

Pilot Participation

The advice of the experienced ship pilots, although limited, was invaluable in validating the suitability of visual displays and the behavior of known vessels under known conditions.

Despite the significant cost savings achieved, the simulator apparently was not used to its fullest capabilities as a design optimization tool. Additional simulations with experienced pilots might have indicated ways to narrow the design width of the connecting channel (for example, by flaring the upper end, which is critical for setting up the turn). The impact of tug usage was not explored until the value engineering study, due to earlier miscommunications with the pilots.

Consensus Building

The simulation study was reported to be of great value in achieving consensus among various interested groups with divergent views on design requirements.

Graphics Displays

Precise fidelity of the graphic display is not always necessary for a successful simulation, but accuracy of perception is. Correct relative placement of visual objects, navigation channel, and environmental data (for example, currents, winds) is critical. The relative importance of display fidelity varies from project to project.

CASE STUDY
PANAMA CANAL GAILLARD CUT WIDENING STUDY
COMPUTER AIDED OPERATIONS RESEARCH
FACILITY, 1983-1986

PROJECT DESCRIPTION

The Panama Canal Commission (PCC) conducted a study of canal modifications to permit two-way traffic of Panamax-size vessels throughout its length which in turn would lead to increased throughput of large ships. At present, the Gaillard Cut is the narrowest section of the canal (Figure C-4). It is 500 feet wide with several curves, which makes the meeting of Panamax vessels dangerous at this point.

The Gaillard Cut Widening Study was intended to determine the dimensions for optimum navigation channel that would afford a reasonable balance between excavation cost and safety. Technical, operational, economic, financial, and environmental considerations were evaluated by the PCC during the study.

SIMULATION DESCRIPTION

The simulation, conducted by the Computer Aided Operations Research Facility (CAORF), was the longest waterway design simulation effort undertaken. Real-time and fast-time simulation were used. Field visits were made to develop visual effects, observe vessels using the canal, and mea-

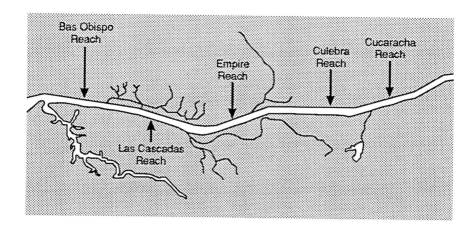

FIGURE C-4 Gaillard Cut, narrowest section of the Panama Canal.

sure environmental effects for simulation. The model ship was developed by the Swedish State Shipbuilding Tank (SSPA) (Eda et al., 1986; Kaufman, 1986; Moody, 1970; Puglisi, 1986; Puglisi et al., 1984, 1987; U.S. Maritime Administration, 1983-1986, 1986).

Criteria for Measuring Safety Performance

Safety criteria were not available for evaluating ship performance in passing maneuvers relative to alternate channel configurations. Therefore, it was necessary to establish the framework and measures needed to evaluate simulation results, including collection of supporting data. A multi-dimensional performance measure, referred to as the steering quality profile, was developed for quantifying the degree of safety achieved during passing evolutions. Meetings and passings of the largest ships currently allowed to meet and pass was selected as the baseline for evaluating comparative performance of Panamax ships, the largest allowed to meet and pass in the widened cut. A generic bulk carrier, 608 feet in length with an 85 foot beam, was selected to represent the largest ships currently using the cut.

Model Validation

The mathematical model was validated using comparisons of simulated ship trajectories and actual ship trajectory data collected during transits of the Gaillard Cut. Subjective evaluations of simulator performance by pilots from the Panama Canal Commission were used to further refine the autopilot model. Simulation scenarios were constructed so that meetings and passings would occur in a straight reach near each turn. The scenarios were further refined so that meetings and passings would occur at the most difficult locations for maneuvering. If acceptable performance were achieved at the most difficult locations, then equal or better performance could reasonably be expected elsewhere.

Fast-time simulation was selected to screen hundreds of design variations and eliminate those that were clearly unacceptable. The coefficients used in mathematical models by SSPA and CAORF for fast-time simulation were derived from SSPA model tests for Panamax ships. Comparison of trajectories from the model test and fast-time simulations using the mathematical models determined that performance agreed within ten percent. This level of accuracy was considered acceptable for initial screening of configuration feasibility.

Fast-time Screening of Design Alternatives

A screening strategy was developed to permit ranking and selection of alternate waterway configurations. Computer data bases were created for over 1500 configurations. Excavation volumes, determined by the commission for each alternative, were used to rank each in terms of economic desirability. The data bases were organized and stored electronically to facilitate data searches.

For each curve in the waterway, operating conditions were correlated with alternate configurations to form a problem matrix. Fast-time screening began with the least costly configurations. Meeting and passing evolutions were executed under varying operating conditions. Operating conditions were derived from observations of actual shiphandling and ship behavior and ranked according to their effect on passing maneuvers. Progressively wider configurations were used to establish minimum widths needed to permit safe passage under progressively more difficult operating conditions. Following each run, data were examined using regimes built into the computer program to confirm that test parameters for operating conditions were within prescribed tolerances (that is, combined meeting speed within plus or minus 1 knot of designated speed, and meeting location within 1/2 ship-length of the intended location). If tolerances were exceeded, the computer program automatically adjusted the initial speeds or starting location and repeated the run. If a run passed screening criterion, then the next set of operating conditions were selected for that configuration. If a run failed, then the next configuration was loaded according to the ranking strategy. The process was automated so that all runs for each turn could be executed and analyzed without human intervention. Based on the overall results, the commission choose one configuration for each curve which struck a balance between cost and the range of operating conditions for which safe passages could be expected. This constituted the finite number of configurations that would be further evaluated using real-time simulation.

Configuration Assessment using Real-Time Simulation

Real-time simulation was used to better determine the acceptability of the waterway configurations selected through the screening process for each turn. Pilots employed by the commission participated in full-mission simulation of meetings and passings of Panamax vessels for selected configurations under varying operating conditions. Steering quality profiles generated from pilot directed maneuvers were compared with baseline criteria for measuring safety performance. For 6 of 8 locations for which data had been collected, average performance of the pilot validation group in terms of ship trajectories was better than the baseline safety level. In 7 of the areas simulated, average pilot performance was equal to or better than baseline safety levels. Only in one area were minor configuration refinements needed to achieve acceptable performance.

SIMULATION RESULTS

Simulation indicated that if the Gaillard Cut was widened to various widths, then longer and wider (in beam) ships could safely pass each other, which would increase canal throughput by many vessels per year.

PROJECT IMPLEMENTATION

Political instability in Panama has impeded the project schedule. The Pacific Entrance Channel modifications have been partially implemented and were scheduled for completion by late 1990. Widening of the Gaillard Cut began in early 1992.

LESSONS LEARNED

Technical

The study demonstrated that simulation can be used as a cost-effective tool in the design process. It further indicated that

- subjective measures are important in the project evaluation process;
- fast-time analysis can be used to determine what layout alternatives are most likely to succeed, thereby overcoming complex interactions that might have otherwise prohibited analysis;
- accuracy of simulation mathematical models could be achieved within 10 percent of each other; and
- visual observation of vessel tracks could be used to validate fast- and real-time simulation.

Simulation results coupled with design manual guidelines resulted in identifying areas for which dredging was not needed. This finding resulted in projected cost savings of up to $400 million. Simulation showed that alternate vessel tracks fell both inside and outside of the guidelines of the Permanent International Association of Nautical Congresses.

Project Management

Lessons learned relevant to project management included:

• pilot participation in the real-time simulation could be used to verify the validity of the simulation to the satisfaction of participants,

• all participants in the study wanted considerable confidence in the appropriateness of the selected design vessel, and

• development of the compressed time simulation decision model and automatic execution process was more complex than standard real-time simulation.

CASE STUDY
GRAYS HARBOR, WASHINGTON
WATERWAYS EXPERIMENT STATION, 1986

PROJECT DESCRIPTION

The Port of Grays Harbor, Washington, wished to verify the feasibility of the final design for a major ship channel improvement project. A design was completed for widening and deepening of 24 miles an estuary and bar channel, improvement of a highway bridge fendering system, and replacement of a rail bridge.

The port sponsored a simulation study by the Waterways Experiment Station (WES) to provide additional input for the final project because of

• high project cost,

• increasing size of the primary vessel type (log carriers) used in original design,

• concern over the adequacy of the turning basin,

• uncertainty over the alignment of the channel in relation to proposed modifications of the bridges, and

• continued concern by environmentalists for a sensitive underwater habitat.

SIMULATION DESCRIPTION

Real-time simulation, using Grays Harbor pilots for verification, was conducted using the WES simulator. Extensive environmental information was available for use in the simulation as part of the basic U.S. Army Corps of Engineers (USACE) navigation project on channel width. All four Grays Harbor pilots participated in the study in close coordination with the local Seattle District, USACE (Hewlett and Nguyen, 1987; Waller and Schuldt, n.d.; Whalin, 1986, 1987).

SIMULATION RESULTS

The simulation indicated that dredging requirements could be safely reduced in the 8 mile channel section from South Reach to Cow Point from 400 feet to 350 feet (existing width) with widening at bends only (see Figure C-5). This finding resulted in a reduction of 1 million cubic yards of dredging in an environmentally sensitive reach out of a total of 17 million cubic yards.

The simulation also indicated that larger vessels than initially anticipated could be safely used in this harbor with only slight modifications in channel design. The size of turning basin was also determined to be adequate for the large ship.

PROJECT IMPLEMENTATION

Simulation recommendations have been incorporated into the final project specifications. The project is currently under construction.

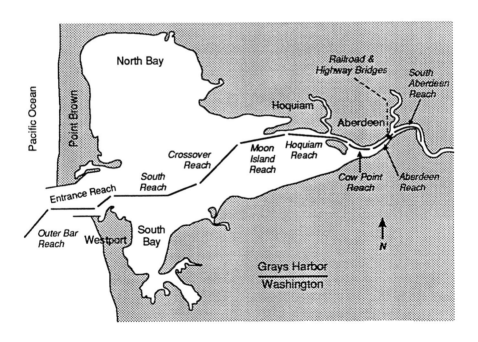

FIGURE C-5   Grays Harbor, Washington.

LESSONS LEARNED

Consensus Building

The simulation made the basic design project credible for a diverse group of project participants. This result was especially valuable when addressing environmental concerns.

Cost Savings Through Simulation

Cost savings that resulted from the Grays Harbor simulation were significant—about 10 times the cost of the simulation.

Design Tool

Current general design rules for ship turning basins may result in construction of basins larger than necessary, based on the Grays Harbor experience.

CASE STUDY
OAKLAND HARBOR
COMPUTER AIDED OPERATIONS RESEARCH
FACILITY, 1986-1988

PROJECT DESCRIPTION

The Port of Oakland sponsored a simulation study at the Computer Aided Operations Research Facility (CAORF) to develop alternative channel designs for the Inner and Outer Oakland harbors (Figure C-6). The objective was to find suitable designs that would open the port to larger, more cost-efficient containerships in order to maintain the port's competitive position relative to other West Coast container ports. A channel dredged to 42 feet was a key project feature. Ships entering the port are subject to adverse current and wind conditions (U.S. Army Corps of Engineers [US-ACE], 1988).

SIMULATION DESCRIPTION

The study used the shiphandling simulator at CAORF to evaluate channel designs. Pilots from the San Francisco Bar Pilots Association participated in the real-time simulation. Two other models were used—the San

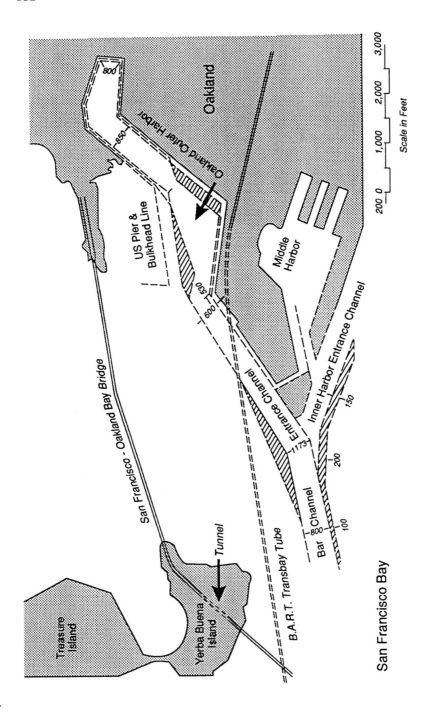

FIGURE C-6   Port of Oakland proposed channel (*USACE*).

Francisco Bay physical model operated by the San Francisco District and the Mooring Line simulation model at the Stevens Institute of Technology. The bay model provided values for water currents. The Mooring Line model determined surge effect on moored vessels. The Waterways Experiment Station (WES) in Vicksburg, Mississippi, provided technical support for current modeling (U.S. Maritime Administration, 1987).

PROJECT RESULTS

The simulation indicated that significant safety benefits could be derived by widening the entrance channel and the westernmost part of the outer harbor channel beyond the widths initially proposed and by tapering the width of the remaining outer channel to its present size, thereby minimizing wake damage to moored vessels. Tapering the outer harbor channel precluded the costly relocations of subway (BART) cables and platforms, which was initially planned. Channel width modifications reduced the turning basin design from 1800 to 1600 feet.

PROJECT IMPLEMENTATION

The final proposed design was approved by the USACE in 1988. The Port of Oakland decided to accelerate phase one of the project (deepening the inner harbor to 38 feet). Dredging was to begin in 1988 as a port-sponsored initiative. However, legal and regulatory challenges concerning port involvement and disposal of dredged material effectively halted the port's implementation activities.

Additional interested parties emerged who were not involved in the simulation or other aspects of the design and approval process. The environmental and commercial fishing concerns of these parties were not satisfied by the disposal plans for dredged material from the prospective project, and implementation was stymied. Some objections, especially from the commercial fishing community, were unexpected because fishing concerns apparently had been addressed. However, not all interested fishermen had been involved, and legal challenges were initiated regarding the offshore disposal of dredged material that they alleged could adversely impact their fishing grounds.

The port overcame or accommodated most of the various challenges to the deepening project. However, the port was unable to obtain approval for a suitable offshore disposal site. Escalating costs that occurred in the interim and additional objections to alternate sites have affected the economic viability of what were once promising disposal options. These options included the possible use of dredged material to restore marine habitats within distant reaches of the estuary and to dispose of it in the Sacramento

River Delta for use as dike construction material by reclamation officials (Kagan, 1990). The port was also unable to resolve legal issues about construction of a part of an authorized federal project by a non-federal organization. As a result, the port discontinued its effort to accelerate completion of phase one and is awaiting USACE implementation following normal procedures.

LESSONS LEARNED

Consensus Building

Simulation brought together parties with interests in project design. It resulted in improved communication and better understanding of project problems, alternatives, and design requirements. Channel design with the aid of simulation provided meaningful data about issues that were previously analyzed subjectively.

Identification of Interested Parties

Although the simulation study was successful, the original identification of parties interested in the project was incomplete, even though some concerns (such as disposal options for dredged material) did not affect design configurations evaluated through simulation. If all parties had been included in the simulation process from the beginning, the same common ground among other parties may not have been achieved. However, the noninvolvement of these late arrivals in the design and approval process has jeopardized project viability (Kagan, 1990).

Design Ship

The appropriate selection of the design ship to be modeled is very important to successful simulation. The Econoline vessel used in this study was not considered to be a good choice. Pilots found it unusually difficult to handle, and it is also believed to be smaller than the ships expected to use the port when the project is completed.

Although additional simulations were recommended to evaluate the performance of the larger vessels expected to use this port (new vessel beam 130 feet versus Econoline beam of 106 feet), the additional work was not done. Thus, the potential exists for larger vessels to operate in a confined, shallow channel without the benefit of supporting simulation data to determine margins of safety for such operations.

Cost Reductions Through Simulation

Simulation can significantly affect the cost-effectiveness and safety of channel designs. This finding was reflected in changes to channel configuration, which reduced overall dredging costs.

D

Source Reference List
for Mathematical Models

Throughout the study, a vast resource of references were identified that could assist practitioners in applying computer-based simulation to channel design. This appendix characterizes in tabular form a representative number of references on mathematical models of system dynamics and force modules in view of their criticality to simulation. Methods for estimating or describing forces and moments on model elements are included only when they define the form of the model structure.

AUTHOR (date)	TITLE	Nature of Treatment or Topic[a] (see note[b] for subject area codes within columns)					
		A	B	C	D	E	F
Abkowitz (1964)	Lectures on Ship Hydrodynamics—Steering and manoeuvrability	1	3				
Abkowitz (1980)	Measurement of Hydrodynamic Characteristics from Ship Manoeuvring Trials by System Identification		1				2
Afremoff and Nikolaev (1972)	Yawing of a Ship Steered by an Automatic Pilot in Rough Seas	6,8					
Ankudinov and Barr (1982)	Estimation of Hydrodynamic Maneuvering Models for Six and Fifteen Barge River Tows		7		7		
Ashburner and Norrbin (1980)	Tug-Assisted Stopping of Large Ships in the Suez Canal—A Study of Safe Handling Techniques		7		7		
Åström et al. (1975)	The Identification of Linear Ship Steering Dynamics Using Maximum Likelihood Parameter Estimation	1	1				1
Baker and Patterson (1969)	Some Recent Developments in Representing Propeller Characteristics		4	4			
Bech (1972)	Some Aspects of the Stability of Automatic Course Control of Ships	1,6	4		1,6		
Bernitsas and Kekridis (1985)	Simulation and Stability of Ship Towing	7			9		
Blanke (1978)	On Identification of Non-Linear Speed Equation from Full-Scale Trials	4					4
Brard (1951)	Maneuvering of Ships in Deep Water, in Shallow Water, and in Canals			2,3			

AUTHOR (date)	TITLE	Nature of Treatment or Topic[a] (see note[b] for subject area codes within columns)					
		A	B	C	D	E	F
Case et al. (1984)	A Comparative Look at the Performance of Simulator Mathematical Models and Future Considerations	9	1				
Cheng et al. (1982)	Flexible Automatic Ship Controllers for Track-Keeping in Restricted Waterways		6		6		
Chislett and Wied (1985)	A Note on the Mathematical Modelling of Ship Manoeuvring in Relation to a Nautical Environment with Particular Reference to Currents		1,8				
Clarke (1971)	A New Non-Linear Equation for Ship Manoeuvring	2	2				
Clarke (1972)	A Two-Dimensional Strip Method for Surface Ship Hull Derivatives: Comparison of Theory with Experiments on a Segmented Tanker Model	2	2	2			
Clarke et al. (1983)	The Application of Manoeuvring Criteria in Hull Design Using Linear Theory	2	2				
Crane (1966)	Studies of Ship Manoeuvring—Response to Propeller and Rudder Actions		4				
Crane, et al (1989)	Controllability	1,9		1,9			1,9
Dand (1975)	Some Aspects of Tug Ship Interaction			3,7			
Dand (1984)	Optimizing Ship Operations in Open and Confined Waters Using Manoeuvring Simulation Models		3				
De Boer (1983)	Manoeuvring Prediction with the MINISIM, A Simulation Program to Predict the Manoeuvring Performance of a Ship		1				

AUTHOR (date)	TITLE	Nature of Treatment or Topic[a] (see note[b] for subject area codes within columns)					
		A	B	C	D	E	F
De Kat and Paulling (1989)	The Simulation of Ship Motions and Capsizing in Severe Seas	1,8		8			
Eda (1967)	Steering Control of Ships in Waves	1,8		8	1,8		
Eda (1972a)	Yaw Control in Waves	1,8		8			
Eda (1972b)	Course Stability, Turning Performance and Connection Force of Barge Systems in Coastal Seawaves	7,8			1		
Eda and Crane (1965)	Steering Characteristics of Ships in Calm Water and Waves	1,8		8	8		
Eda and Savitsky (1969)	Experimental and Analytical Studies of Ship Controllability in Canals		3	3	3		
Edwards (1985)	Hydrodynamic Forces on Vessels Stationed in a Current		8	8			
Eskola (1986)	Modelling the Propulsion Machinery Behavior During Model Propulsion Tests in Ice		5,8				
Forsman and Sandkvist (1986)	Brash Ice Effects on Ship Operations—A Presentation of the SSPA Manoeuvring Simulation Model and Other Brash Ice Related Projects		8				
Fujii (1972)	On Manoeuvre Tests to Investigate the Course-Keeping Qualities of Ships				3		
Fujino (1968)	Experimental Studies on Ship Manoeuvrability in Restricted Waters, Part 1	1,3	3	3			
Fujino (1970)	Experimental Studies on Ship Manoeuvrability in Restricted Waters, Part 2	1,3	3	3			

AUTHOR (date)	TITLE	Nature of Treatment or Topic[a] (see note[b] for subject area codes within columns)					
		A	B	C	D	E	F
Fujino and Ishiguro (1984)	A Study of the Mathematical Model Describing Manoeuvring in Shallow Water—Shallow Water Effects on Rudder Effectiveness Parameters		3,4				
Gertler and Hagen (1967)	Standard Equations of Motion for Submarine Simulation	1,2	2				
Gill (1979)	Mathematical Modelling of Ship Manoeuvring		1				
Glansdorp (1975)	Ship Type Modelling for a Training Simulator		1				
Göransson and Liljenberg (1975)	Simulating the Main Engine—A Comparison of FPP and CPP Arrangements (in Swedish with English Summary)		5		5		
Hagen (1983)	A Catalog of Existing Mathematical Models for Maneuvering	1	1				
Hirano (1980)	On Calculation Method of Ship Manoeuvring Motion at Initial Design Stage (in Japanese)		1,2				
Hirano et al. (1985)	A Computer Program System for Ship Manoeuvring Motion Prediction	8	1		1		
Hoffman (1972)	Consideration of Sea-Keeping in the Design of a Ship Manoeuvring Simulator	1,6				1,8	
Holzhüter (1990)	A Workable Dynamic Model for the Track Control of Ships	1,6		6			
Hooft (1968)	The Manoeuvrability of Ships on a Straight Course	1			1		
Hu (1961)	Forward Speed Effect on Lateral Stability Derivatives of a Ship	2		2			

AUTHOR (date)	TITLE	Nature of Treatment or Topic[a] (see note[b] for subject area codes within columns)					
		A	B	C	D	E	F
Imlay (1961)	The Complete Expressions for Added Mass of a Rigid Body Moving in an Ideal Fluid	2					
Inoue and Murayama (1970)	Calculation of Turning Ship Derivatives in Shallow Water (in Japanese)		3	3			
Inoue et al. (1981)	A Practical Calculation Method of Ship Manoeuvering Motion		1				
Jacobs (1964)	Estimation of Stability Derivatives and Indices of Various Ship Forms, and Comparison with Experimental Results	2	2	2			
Källström (1979)	Identification and Adaptive Control Applied to Ship Steering	6					2
Källström (1984)	A Digital Control System for Ship Manoeuvring in Ports and Waterways	6			6		
Källström and Ottosson (1982)	The Generation and Control of Roll Motion of Ships in Close Turns		1,6		6		
Kijima et al. (1990)	Prediction Method of Ship Manoeuvrability in Deep and Shallow Waters	2,3	3				
Kirchhoff (1869)	Über die Bewegung eines Rotationskörpers in einer Flüssigkeit	1,2	1,4				
Kobayashi (1988)	A Simulation Study on Ship Manoeuvrability at Low Speeds		1,4		1		
Koelink (1968)	Approximate Methods in Z-Steering Test Analysis	1	2				
Kose (1982)	On a New Mathematical Model of Manoeuvring Motions of a Ship and Its Applications	1	1,4		1		
Kotschin et al. (1954)	Theoretische Hydromechanik	2					

AUTHOR (date)	TITLE	Nature of Treatment or Topic[a] (see note[b] for subject area codes within columns)					
		A	B	C	D	E	F
Koyama (1972)	Improvement of Course Stability and Control by a Subsidiary Automatic Control	1,6			6		
Koyama and Jin (1987)	An Expert System Approach to Collision Avoidance	1,6					
Koyama et al. (1977)	A Study of the Instability Criterion on the Manual Steering of Ships		6		6		
Lamb (1918)	The Inertia Coefficients of an Ellipsoid Moving in a Fluid	2		2			
Landweber and de Macagno (1957)	Added Mass of Two-Dimensional Forms Oscillating in a Free Surface	2		2			
Lindström (1989)	Prediction of Ship Manoeuvring in Level Ice by Simulation of the Planar Motions (in Swedish)		8		8		
Mandel (1967)	Ship Maneuvering and Control	1,9		1,9			
Matsumoto and Suemitsu (1984)	Interference Effects Between Hull, Propeller and Rudder of a Hydrodynamic Mathematical Model in Manoeuvring Motion		2	4			
Matthews (1984)	A Six Degree of Freedom Ship Model for Computer Simulation	2	1				
McCallum (1976)	A New Approach to Manoeuvring Ship Simulation	2	1				
McCallum (1980)	A Ship Steering Mathematical Model for All Manoeuvring Regimes		1,4				
McCreight (1986)	Ship Maneuvering in Waves	1,8			8		
Mikelis et al. (1985)	On the Construction of a Versatile Mathematical Model for Marine Simulation		1,2				

AUTHOR (date)	TITLE	Nature of Treatment or Topic[a] (see note[b] for subject area codes within columns)					
		A	B	C	D	E	F
Miller (1979)	Towboat Maneuvering Simulator		1,4 7				
Motora (1960)	On the Measurement of Added Mass and Added Moment of Inertia of Ships in Steering Motion			2			
Motora et al. (1971)	Equivalent Added Mass of Ships in the Collision			2			
Naegle (1980)	Ice Resistance Prediction and Motion Simulation for Ships Operating in the Continuous Mode of Icebreaking		8		8		
Newman (1966)	Some Hydrodynamic Aspects of Ship Maneuverability	1,2		2			
Newman (1969)	Lateral Motion of a Slender body of Revolution Moving near a Wall	3					
Newman (1972)	Some Theories for Ship Manoeuvring	3					
Nikolaev et al. (1972)	Estimation of the Effectiveness of Lateral Thrust Units			4			
Nomoto (1960)	Analysis of Kempf's Standard Manoeuvre Test and Proposed Steering Quality Indices		1				1
Nomoto (1966)	Unusual Scale Effects on Manoeuverabilities of Ships with Blunt Bodies			2			2
Nomoto (1972)	Problems and Requirements of Directional Stability and Control of Surface Ships	1	1				
Nomoto et al. (1957)	On the Steering Qualities of Ships	1	1				
Norrbin (1960)	A Study of Course Keeping and Manoeuvring Performance	1,6	9				

AUTHOR (date)	TITLE	Nature of Treatment or Topic[a] (see note[b] for subject area codes within columns)					
		A	B	C	D	E	F
Norrbin (1963)	On the Design and Analysis of the Zig Zag Test on Base of Quasi-Linear Frequency Response	1	2				
Norrbin (1965)	The Technique and Analysis of the Zig Zag Test (in Swedish)	1	1				1
Norrbin (1970)	Theory and Observations on the Use of a Mathematical Model for Ship Manoeuvring in Deep and Confined Waters	1,2	2,6	2,3		1,6	
Norrbin (1972)	Ship Manoeuvring with Application to Shipborne Predictors and Real-Time Simulators		1,6				
Norrbin (1978)	A Method for the Prediction of the Manoeuvring Land of a Ship in a Channel of Varying Width		2,3	3			
Norrbin (1986)	Fairway Design with Respect to Ship Dynamics and Operational Requirements		3,6 7	3	3		
Norrbin (1988)	Head-On Collision or a Planned Encounter—A Contribution to Micro-Navigation		3	3	6		
Norrbin et al. (1978)	A Study of the Safety of Two-Way Traffic in a Panama Canal Bend	1	2,3	3	1,6	1	
Ogawa and Kasai (1978)	On the Mathematical Model of Manoeuvring Motion of Ships	1	1				
Oltmann and Sharma (1985)	Simulation of Combined Engine and Rudder Manoeuvres Using an Improved Model of Hull-Propeller-Rudder Interactions	1	1,4				
Onassis and ten Hove (1988)	Modular Ship Manoeuvring Models		1,4				

AUTHOR (date)	TITLE	Nature of Treatment or Topic[a] (see note[b] for subject area codes within columns)					
		A	B	C	D	E	F
Ottosson and Apleberger (1987)	Real-Time Simulations of Ship Motions—A Tool for the Design of a New Pacific Port		7,8			1,3	1,3
Ottosson and van Berlekom (1985)	Simulations in Real and Accelerated Time—A Computer Study of a Ro-Ro Vessel Entering Different Port configurations				1	1	
Paloubis and Thaler (1972)	Identification of System Models from Operating Data	1					1,5
Perdok and van der Tak (1987)	The Application of Man-Machine Models in the Analysis of Ship Control	6	6				6
Perez y Perez (1972)	A Time Domain Solution to the Motions of a Steered Ship in Waves	1,8	8		1,8		
Pourzanjani (1990)	Formulation of the Force Mathematical Model of Ship Manoeuvring	1	2				2
Pourzanjani et al. (1987)	Hydrodynamic Lift and Drag Simulation for Ship Manoeuvring Models		2				
Price (1972)	The Stability of a Ship in a Simple Sinusoidal Wave	1,8					
Puglisi et al. (1985a)	Direction of International Joint Effort for Development of Mathematical Models and Ship Performance Data for Marine Simulation Application	9					1
Renilsson and Driscoll (1982)	Broaching—An Investigation into the Loss of Directional Control in Severe Following Seas		8				
Rutgersson and Ottosson (1987)	Model Tests and Computer Simulations—An Effective Combination for Investigation of Broaching Phenomena	8			1,8		

AUTHOR (date)	TITLE	Nature of Treatment or Topic[a] (see note[b] for subject area codes within columns)					
		A	B	C	D	E	F
Rydill (1959)	A Linear Theory for the Steered Motion of Ships in Waves	1,6			1		
Salo and Heikkilä (1990)	On the Modelling of Hull-Propeller-Rudder Interactions in Manoeuvring of Twin-Screw ships		4				4
Sargent and Kaplan (1970)	System Identification of Surface Ship Dynamics						2
Schmidt and Unterreiner (1976)	Ein Mathematisches Modell zur Simulation des Manövrierverhaltens von Schiffen für die Anwendung in Trainings-Simulatoren		1				
Schoenherr (1960)	Data for Estimating Bank Suction Effects in Restricted Water and on Merchant Ship Hulls			3			
Shooman (1980)	Models of Helmsman and Pilot Behavior for Manoeuvring Ships				6		
Smitt (1970)	Steering and Manoeuvring: Full-Scale and Model Tests	1	2	2			
Smitt and Chislett (1972)	Course Stability While Stopping	2	2,4	4			
Society of Naval Architects and Marine Engineers (1950)	Nomenclature for Treating the Motion of a Submerged Body Through a Fluid	1,2	2				
Strøm-Tejsen (1965)	A Digital Computer Technique for Prediction of Standard Maneuvers of Surface Ships		1,2				

AUTHOR (date)	TITLE	Nature of Treatment or Topic[a] (see note[b] for subject area codes within columns)					
		A	B	C	D	E	F
Strøm-Tejsen and Chislett (1966)	A Model Testing Technique and Method of Analysis for the Prediction of Steering and Manoeuvring Qualities of Surface Ships		2	2			
Stuurman (1969)	Modelling the Helmsman: A Study to Define a Mathematical Model Describing the Behavior of a Helmsman Steering a Ship Along a Straight Course		6		6		
Tasai (1961)	Hydrodynamic Force and Moment Produced by Swaying Oscillation of Cylinders on the Surface of a Fluid	2	6	2			
Thöm (1975)	Modellbildung für das Kursverhalten von Schiffen	6	6				
Trägårdh (1976)	Simulation of Tugs at the SSPA Manoeuvring Simulator	3	7				
Tuck (1966)	Shallow Water Flows Past Slender Bodies	3		3			
Tuck and Newman (1974)	Hydrodynamic Interactions Between Ships	3		3			
Van Amerongen and van der Klugt (1985)	Modelling and Simulation of the Roll Motions of a Ship	1			1		
Van Berlekom (1978)	Simulator Investigations of Predictor Steering Systems for Ships	1,6		2	6		
Van Berlekom and Goddard (1972)	Maneuvering of Large Tankers		1	2			
Van Leeuwen (1964)	The Lateral Damping and Added Mass of an Oscillating Shipmodel			2			

AUTHOR (date)	TITLE	Nature of Treatment or Topic[a] (see note[b] for subject area codes within columns)					
		A	B	C	D	E	F
Van Leeuwen (1972a)	Some Aspects of Prediction and Simulation of Manoeuvres		1		1		
Van Leeuwen (1972b)	Course Keeping Going Astern		2,4	2,4			
Veldhuijzen and Stassen (1975)	Simulation of Ship Manoeuvring Under Human Control		6		1		
Vugts (1968)	The Hydrodynamic Coefficients for Swaying, Heaving and Rolling Cylinders in a Free Surface	2		2			
Webster (1967)	Analysis of the Control of Activated Antiroll Tanks	4,8					
Weinblum (1952)	On the Directional Stability of Ships in Calm Water and in a Regular Seaway	1,8					
Wendel and Dunne (1969)	Dynamic Analysis and Simulation of Ship and Propulsion Plant Manoeuvring Performance		4				
Yeung (1978)	Applications of Slender Body Theory to Ships Moving in Restricted Shallow Water	3		3			
Zhao (1990)	Theoretical Determination of Ship Manoeuvring Motion in Shallow Water	3					
Zhou and Blanke (1987)	Nonlinear Recursive Prediction Error Method Applied to Identification of Ship Steering Dynamics	1					1
Zuidweg (1970)	Automatic Guidance of Ships as a Control Problem	6	6				

^aCodes for nature of treatment or topic:

A   Theory and theoretical models
B   Semiempirical models
C   Data figures
D   Applications in compressed time
E   Applications in real time
F   Identification, validation

^bCodes for subject areas:

1   System dynamics
2   Hull forces in deep water
3   Hull forces in confined water (shallow, restricted)
4   Control forces (propulsive, lateral)
5   Engine functioning (propulsive and lateral control)
6   Control automatics, human pilots (including fast time/compressed simulation)
7   Tug assistance, mooring, fendering
8   External forces from wind, waves, current, mud, ice
9   General or nonspecific

E

Papers Prepared for This Study

Barr, R. 1991. Comparison of simulation models and validation studies for the *Esso Osaka*. Background paper prepared for the Committee on Assessment of Shiphandling Simulation, National Research Council, Washington, D.C.

Norrbin, N. H. 1991. Source reference list on mathematical models. Background paper prepared for the Committee on Assessment of Shiphandling Simulation, National Research Council, Washington, D.C.

Norrbin, N. H., R. A. Barr, C. L. Crane, Jr., and W. C. Webster. 1989. A summary of findings from task group visits to a number of european ship manoeuvreing simulator centers. Background paper prepared for the Committee on Assessment of Shiphandling Simulation, National Research Council, Washington, D.C.

Puglisi, J. J. 1989. European trip report. Background paper prepared for the Committee on Assessment of Shiphandling Simulation, National Research Council, Washington, D.C.

F

Validation of Aircraft Flight Simulators

OVERVIEW

Computer-based simulations are used widely in commercial aviation to assist in airframe design, flight operations, and pilot training (Stix, 1991). Development of aircraft flight simulators is directly linked to development of specific aircraft. The extensive data generated as part of aircraft design and testing are used as a technical resource for developing a simulator for training pilots in that aircraft's operation (aircraft flight simulators are airframe-specific). Thus, aircraft flight simulators cannot be modified to permit training in multiple airframes nor are they used for designing air routes.

Validation of an aircraft flight training simulator's fidelity to represent an aircraft's performance and handling historically has been the task of the chief pilot for an airframe manufacturer, a pilot selected by the Federal Aviation Administration (FAA), or a military officer assigned the role of project pilot for the Department of Defense. Evaluations have been based on a subjective opinion of the pilot relative to how well the cockpit controllers (such as stick, throttles, rudder pedal, and brakes), cockpit instrumentation, aural system (used to generate engine sounds and wind noise), visual system, and motion system are designed, modeled, and integrated to recreate the true behavior of the aircraft for various flight mission phases (for example, ground handling, takeoff, and climb). Today, pilots still play a role in the simulation validation process. However, many quantitative tests have been designed and used for evaluating the correctness of the simulator.

VALIDATION POLICY

Strict guidelines are followed for the design and validation of military operational flight and weapon system trainers. The military software specification MIL-2167A defines the procedure by which simulator software is designed, documented, and validated. Simulators for aircraft under the jurisdiction of the FAA are validated under criteria specified by an FAA Advisory Circular. Currently, this is AC120-45A Draft: *Airplane Flight Training Device Qualification*. The FAA's interest in certificating an aircraft flight simulator can be traced to its philosophy concerning the role of flight simulators. This viewpoint is stated in the introduction to the circular, which follows.

The primary objective of flight training is to provide a means for flight crewmembers to acquire the skills and knowledge necessary to perform to a desired safe standard. Flight simulation provides an effective, viable environment for the instruction, demonstration, and practice of the maneuvers and procedures (called training events) pertinent to a particular airplane and crewmember position. Successful completion of flight training is validated by appropriate testing, called checking events. The complexity, operating costs, and operating environment of modern airplanes, together with the technological advances made in flight simulation, have encouraged the expanded use of training devices and simulators in the training and checking of flight crewmembers. These devices provide more in-depth training than can be accomplished in the airplane and provide a very high transfer of skills, knowledge, and behavior to the cockpit. Additionally, their use results in safer flight training and cost reductions for the operators, while achieving fuel conservation, a decrease in noise and otherwise helping maintain environmental quality.

The FAA has traditionally recognized the value of training devices and has awarded credit for their use in the completion of specific training and checking events in both general aviation and air carrier flight training programs and in pilot certification activities. Such credits are delineated in FAR Part 61 and Appendix A of that part; FAR Part 121, including Appendices E and F; and in other appropriate sources such as handbooks and guidance documents. These FAR sources, however, refer only to a "training device," with no further descriptive information. Other sources refer to training devices in several categories such as Cockpit Procedures Trainers (CPT), Cockpit Systems Simulators (CSS), Fixed Base Simulators (FBS), and other descriptors. These categories and names have had no standard definition or design criteria within the industry and, consequently, have presented communications difficulties and inconsistent standardization in their application. Furthermore, no single source guidance document has existed to categorize these devices, to provide qualification standards for each category, or to relate one category to another in terms of capability or technical complexity. As a result, approval of these devices for use in training programs has not always been equitable.

The circular, under Evaluation Policy, addresses the scope of quantification testing that is required in order to certificate (validate) the operation of a simulator, as follows:

> The flight training device must be assessed in those areas which are essential to accomplishing responses and control checks, as well as performance in the takeoff, climb, cruise, descent, approach, and landing phases of flight. Crewmember station checks, instructor station functions checks, and certain additional requirements depending on the complexity of the device (i.e., touch activated, cathode ray tube instructor controls; automatic lesson plan operation; selected mode of operation for "fly-by-wire" airplanes; etc.) must be thoroughly assessed. Should a motion system or visual system be contemplated for installation on any level of flight training device, the operator or the manufacturer should contact the NSPM for information regarding an acceptable method for measuring motion and/or visual system operation and application tolerances. The motion and visual systems, if installed, will be evaluated to ensure their proper operation.

> The intent is to evaluate flight training devices as objectively as possible. Pilot acceptance, however, is also an important consideration. Therefore, the device will be subjected to the validation tests listed in Appendix 2 of this Advisory Circular and the functions and subjective tests from Appendix 3. These include a qualitative assessment by an FAA pilot who is qualified in the respective airplane, or set of airplanes in the case of Level 2 or 3. Validation tests are used to compare objectively flight training device data and airplane data (or other approved reference data) to assure that they agree within a specified tolerance. Functions tests provide a basis for evaluating flight training device capability to perform over a typical training period and to verify correct operation of the controls, instruments, and systems.

QUANTITATIVE TEST PROCEDURES

Systematic procedures have been and are continuing to be developed to aid the validation of all components that comprise a modern aircraft simulator. Parameter identification is now being used routinely to extract aerodynamic models from flight test measurements. The parameter identification results are used to validate the simulation mathematical model (Anderson et al., 1983; Anderson and Vincent, 1983; Hess and Hildreth, 1990; Trankle et al., 1981; Trankle et al., n.d.).

Validation of an aircraft flight simulation model typically involves four levels of testing (actual procedures vary by facility). At the first level, individual modules or sub-programs (down to the smallest practical subdivision) are tested individually. This insures that each module has been coded correctly (that is, it satisfies the design requirements for that module). The second level involves testing of small program packages or groups of sub-programs that are related in functionality (such as those modules that comprise the propulsion system). These are tested as separate packages to further debug them and to test the input/output interaction between each module. Model validation also begins in its simplest form.

Test drivers are used in both the first and second levels. For example, a test driver has been developed for static testing of subroutines by allowing control of the input and output variables to each group of program packages. Inputs are generated to stimulate each individual program package in a controlled manner so that the resulting outputs can be examined, usually graphically. For example, in aero models, the angle-of-attack is varied at −180° to +180° for given Mach numbers. The coefficients that comprise the aero model are then plotted as a function of angle-of-attack to assure that their value is correct and continuous. Testing at the second level completes the static testing of the math model.

The complete math model is tested for dynamic response verification at level three. Two test tools are used for analysis. An open loop test generator generates step, sine wave, or ramp or doublet inputs to the simulation. These are used to assess dynamic responses. Since the inputs are computer generated, they can be reproduced exactly and can be used to produce easily analyzed inputs. A second program is used to drive a simulator with aircraft flight test data. This program has the capability of over-driving aircraft math model states or controls with those of the aircraft. For instance, the flight control system can be completely validated by using the flight test feedbacks such as pitch rate and normal acceleration and other measures as inputs to the flight controls along with input from the pilot. This way, the outputs of the flight controls such as control surface positions can be examined on a one-for-one basis with the output of the flight controls in the aircraft. If the simulator flight controls have the same inputs as the aircraft flight controls, their control surface deflection should be the same. Likewise, the aero response of the simulator can be isolated from the effect of the flight controls. This can be done by driving the aero simulation with the actual control surface deflections recorded in the flight test program and then examining the dynamic response of the simulator compared to the aircraft. The same procedure can be followed for engine validation.

Final testing, consisting of pilot-in-the-loop, is performed at level four. By this time, the math model has already been validated but adjustments may have to be made to gain pilot approval. These adjustments primarily result from the limited ability of the motion and visual system to realistically simulate pilot cues. Adjustments determined necessary are satisfied by improving cuing system compensation.

Bibliography

Abkowitz, M. A. 1964. Lectures on Ship Hydrodynamics—Steering and Manoeuverability. Hydo- and Aerodynamics Laboratory, Report Number Hy-5. May. Copenhagen, Denmark: Danish Technical Press.

Abkowitz, M. A. 1980. Measurement of hydrodynamic characteristics from ship manoeuvering trials by system identification. Society of Naval Architects and Marine Engineers (SNAME) Transactions 88:283-318.

Abkowitz, M. A. 1984. Measurement of Ship Hydrodynamic Coefficients in Maneuvering from Simple Trials During Regular Operation. Massachusetts Institute of Technology (MIT) Report MIT-OE-84-1. November. Cambridge, Mass: MIT.

Afremoff, A. Sh., and E. P. Nikolaev. 1972. Yawing of a ship steered by an automatic pilot in rough seas. Journal of Mechanical Engineering Science 14(7)(suppl.):210-270.

Anderson, L., and J. H. Vincent. 1983. AV-8B System Identification Results from Full-Scale Development Flight Test Program (NPE-1). Naval Air Test Center Technical Memorandum, December. Patuxent, Md.: Naval Air Test Center.

Anderson, L., J. H. Vincent, and B. L. Hildreth. 1983. AV-8B System Identification Results from Full-Scale Development Flight Test Program. American Institute of Aeronautics and Astronautics (AIAA) Paper No. AIAA Paper 83-2746 presented at the AIAA Second Flight Testing Conference, Las Vegas, Nev., November 16-18, 1983.

Ankudinov, V., and R. Barr. 1982. Estimation of Hydrodynamic Maneuvering Models for Six and Fifteen Barge River Tows. Technical report 83011-1. Laurel, Md.: Hydronautics.

Ankudinov, V., L. Daggett, C. Hewlett, B. Jakobsen, and E. Miller, Jr. 1989. Use of Simulators in Harbor and Waterway Development. Pp. 267-282 in K. M. Childs, Jr., ed., Ports '89. New York: American Society of Civil Engineers.

Ankudinov, V., and E. Miller. 1977. Predicted Maneuvering Characteristics of the Tanker Esso Osaka in Deep and Shallow Water. Laurel, Md.: Hydronautics.

Armstrong, M. C. 1980. Practical Shiphandling. Glasgow, Scotland: Brown, Son and Ferguson.

Ashburner, R. H., and N. H. Norrbin. 1980. Tug-assisted stopping of large ships in the Suez Canal—A study of safe handling techniques. Pp. 14.1-14.45 in Proceedings of the International Symposium on Ocean Engineering/Ship Handling. Gothenburg, Sweden: Swedish Maritime Research Centre.

Astle, P., and L. H. van Opstal. 1990. ECMAN: A standard database management system for electronic chart display and information systems (ECDIS). International Hydrographic Review 67:101-105.

Åström, K. J., C. G. Källström, N. H. Norrbin, and L. Byström. 1975. The Identification of Linear Ship Steering Dynamics Using Maximum Likelihood Parameter Estimation. Swedish State Shipbuilding Tank (SSPA) Publ. No. 75. Gothenburg, Sweden: SSPA Maritime Consulting.

Atkins, D. A., and W. R. Bertsche. 1980. Evaluation of the safety of ship navigation in harbors. Paper presented to Society of Naval Architects and Marine Engineers, Spring Meeting, STAR Symposium, Coronado, Calif., June 4-8, 1980.

Baker, D. W., and C. L. Patterson, Jr. 1969. Some recent developments in representing propeller characteristics. Pp. AP-A-1–AP-A-34 in Proceedings of the Second Ship Control Systems Symposium, Annapolis, Md. Annapolis, Md.: Naval Ship Research and Development Laboratory.

Bech, M. I. 1972. Some aspects of the stability of automatic course control of ships. Journal of Mechanical Engineering Science 14(7)(suppl.):123-131.

Bernitsas, K. M. M., and N. S. Kekridis. 1985. Simulation and stability of ship towing. International Shipbuilding Progress 32(369):112-123.

Bertsche, W., and R. C. Cook. 1980. A systematic approach for evaluation of port development and operations problems utilizing real time simulation. Pp. D1-01–D1-11 in Proceedings, Fourth Annual CAORF Symposium: Ship Operations Research for the Maritime Industry—Looking to the 80's. Kings Point, N.Y.: National Maritime Research Center.

Blanke, M. 1978. On identification of non-linear speed equation from full scale trials. Paper No. 3 in Proceedings of Fifth Ship Control Systems Symposium, vol. 6. Bethesda, Md.: David Taylor Naval Ship Research and Development Center.

Bogdonov, P., P. Vassilev, M. Leflerova, and E. Milanov. 1987. *Esso Osaka* tanker manoeuvrability investigations in deep and shallow water using PMM. International Shipbuilding Progress 34(390):30-39.

Bowditch, N. 1981. American Practical Navigator: Useful Tables, Calculations, Glossary of Marine Navigation, vol. 2. H.O. Publication No. 9. Washington, D.C.: Defense Mapping Agency Hydrographic/Topographic Center.

Brady, E. M. 1967. Tugs, Towboats and Towing. Cambridge, Md.: Cornell Maritime Press.

Brard, R. 1951. Maneuvering of ships in deep water, in shallow water, and in canals. Society of Naval Architects and Marine Engineers (SNAME) Transactions 59:229-257.

Brown, W., F. E. Guest, W-Y. Hwang, and C. Pissariello. 1989. Factors to Consider in Developing a Knowledge-Based Autopilot Expert System for Ship Maneuvering Simulation. Paper presented at Spring Meeting/STAR Symposium, New Orleans, La, April 12-15, 1989. Jersey City, N.J.: Society of Naval Architects and Marine Engineers.

Bruun, P. 1989a. Port Engineering, vol. 1: Harbor Planning, Breakwaters, and Marine Terminals. Houston, Texas: Gulf Publishing Company.

Bruun, P. 1989b. Port Engineering, vol. 2: Harbor Transportation, Fishing Ports, Sediment Transport, Geomorphology, Inlets, and Dredging. Houston, Texas: Gulf Publishing Company.

Burgers, A., and M. Kok. 1988. The Statistical Analysis of Ship Manoeuvreing Simulator

Results for Fairway Design Based on the Interdependency of Fairway Cross-Section Transits. Pp. 1.123-1.135 in Proceedings of the Ninth International Harbour Conference. Antwerp, Belgium: Royal Society of Flemish Engineers.

Burgers, A., and G. J. A. Loman. 1985. Statistical Treatment of Ship Manoeuvreing Results for Fairway Design. Permanent International Association of Navigation Congresses (PIANC) Bulletin No. 45. Brussels, Belgium: PIANC.

Case, J. S., J. B. van der Brug, and J. J. Puglisi. 1984. A comparative look at the performance of simulator mathematical models and future considerations. Pp. 399-412 in Proceedings of the Third International Conference of Marine Simulation (MARSIM 84). Rotterdam, Netherlands: Maritime Research Institute Netherlands.

Case, J. S., H. Eda, and P. K. Schizume. 1985. The Development and Validation of Hydrodynamic Models. Section A4 in J. J. Puglisi, ed., Proceedings of Sixth CAORF Symposium. Kings Point, N.Y.: National Maritime Research Center.

Cheng, R. C. H., J. E. Williams, and I. R. McCallum. 1982. Flexible automatic ship controllers for track-keeping in restricted waterways. In Proceedings of Fourth International Federation for Information Processing (IFIP)/International Federation of Automatic Control (IFAC) Symposium. New York: North Holland Publishing Company.

Chislett, M. S., and S. Wied. 1985. A Note on the Mathematical Modelling of Ship Manoeuvreing in Relation to a Nautical Environment with Particular Reference to Currents. Pp. 119-129 in Proceedings of the International Conference on Numerical and Hydraulic Modelling of Ports and Harbours, Birmingham, England. Cranfield, Bedford, England: British Hydrodynamics Research Associates.

Clarke, D. 1971. A new non-linear equation for ship manoeuvreing. International Shipbuilding Progress 18(201):181-197.

Clarke, D. 1972. A two-dimensional strip method for surface ship hull derivatives: Comparison of theory with experiments on a segmented tanker model. Journal of Mechanical Science 14(7)(suppl.):53-61.

Clarke, D. 1990. Visions of a super-automated future. The Naval Architect April:E182, E184.

Clarke, D., P. Gedling, and G. Hine. 1983. The application of manoeuvreing criteria in hull design using linear theory. Transactions of the Royal Institute of Naval Architects (RINA) 125:45-68.

Converse, H., D. Hancock, G. Otoshi, and H. Erlich. 1987. Ship simulation modeling: Experiences and potential within the South Pacific Division. Pp. 21-41 in Proceedings of San Francisco District Navigation Workshop, May 18-21, 1987. San Francisco, Calif.: U.S. Army Corps of Engineers District, San Francisco.

Crane, C. L., Jr. 1966. Studies of ship manoeuvreing—Response to propeller and rudder actions. Supplemental paper in Proceedings of Ship Control Systems Symposium, November 13-15, 1966. Annapolis, Md.: U.S. Navy Marine Engineering Laboratory.

Crane, C. L., Jr. 1979a. Maneuvering trials of the 278,000 DWT *Esso Osaka* in shallow and deep water. Society of Naval Architects and Marine Engineers (SNAME) Transactions 87:251-283.

Crane, C. L., Jr. 1979b. Maneuvering Trials of the 278,000 DWT *Esso Osaka* in Shallow and Deep Water. EXXON International Report No. EII.4TM.79. Florham Park, N.J.: Exxon Company, International.

Crane, C. L., Jr., H. Eda, and A. Landsburg. 1989. Controllability. Pp. 191-429 in E. V. Lewis, ed., Principles of Naval Architecture (2d rev.), vol. 3. Jersey City, N.J.: Society of Naval Architects and Marine Engineers.

Crenshaw, R. S., Jr. 1975. Naval Shiphandling. Annapolis, Md.: Naval Institute Press.

Dand, I. W. 1975. Some Aspects of Tug Ship Interaction. Paper A5 in Proceedings of the Fourth International Tug Convention, New Orleans, La. London, England: Ship and Boat International.

Dand, I. W. 1981. An Approach to the Design of Navigation Channels. Report R104, Project 252001. Feltham, England: National Maritime Institute.

Dand, I. W. 1984. Optimizing ship operations in open and confined waters using manoeuvreing simulation models. In Proceedings of the West European Conference on Marine Technology (WEMT '84), Paris. Paris, France: Association Technique et Maritime.

Dand, I. W. 1987. On modular manoeuvreing models. Paper No. 8 in International Conference on Ship Manoeuverability: Prediction and Achievement, London, April 29-30 and May 1, 1987, vol. 1. London, England: Royal Institute of Naval Architects.

Dand, I. W. 1988. Hydrodynamic aspects of the sinking of the ferry *Herald of Free Enterprise*. Paper presented at a meeting of the Royal Institute of Naval Architects, April 20, 1988. London, England.

Dand, I. W. 1989. Discussion document on error analysis in manoeuvreing studies. Paper prepared for Nineteenth International Towing Tank Conference (ITTC) Manoeuvrability Committee, January. Middlesex, England: British Maritime Technology.

Dand, I. W., and D. B. Hood. 1983. Manoeuvreing Experiments Using Two Geosims of the *Esso Osaka*. National Maritime Institute (NMI) Report No. NMI R 163. Feltham, England: NMI.

De Boer, W. 1983. Manoeuvreing Prediction with the MINISIM, A Simulation Program to Predict the Manoeuvreing Performance of a Ship. Report no. 596-M. Delft, Netherlands: Delft University of Technology.

De Kat, J. O., and J. R. Paulling. 1989. The simulation of ship motions and capsizing in severe seas. Society of Naval Architects and Marine Engineers (SNAME) Transactions 97:139-168.

DeLoach, S. R. 1990. Decimeter positioning and navigation with the global positioning system. Sea Technology (August):23-27.

Eaton, R. M., H. Astle, S. J. Glavin, S. T. Grant, S. E. Masry, and B. W. Shaw. 1990. Learning from an electronic chart testbed. International Hydrographic Review 67(July):31-43.

Eda, H. 1967. Steering Control of Ships in Waves. Davidson Laboratory Report 1205. Hoboken, N.J.: Davidson Laboratory.

Eda, H. 1971. Directional stability and control of ships in restricted channels. Society of Naval Architects and Marine Engineers (SNAME) Transactions 79:71-116.

Eda, H. 1972a. Yaw control in waves. Journal of Mechanical Engineering Science 14(7)(suppl.):211-215.

Eda, H. 1972b. Course stability, turning performance and connection force of barge systems in coastal seawaves. Society of Naval Architects and Marine Engineers (SNAME) Transactions 80:299-328.

Eda, H. 1973. Dynamic Behavior of Tankers During Two-Way Traffic in Channels. Marine Technology 10:229-250.

Eda, H. 1979a. Development and Verification of Mathematical Models for Shiphandling Simulators. Pp. 18-1 to 18-10 in Proceedings, Third Annual Computer Aided Operations Research Facility (CAORF) Symposium, Maritime Simulation Research. Kings Point, N.Y.: National Maritime Research Center.

Eda, H. 1979b. Studies on Vessel Maneuvering Characteristics—Shallow Water Effects. Davidson Laboratory Report SIT-DL-79-9-2012. Hoboken, N.J.: Davidson Laboratory.

Eda, H., and C. L. Crane, Jr. 1965. Steering characteristics of ships in calm water and waves. Society of Naval Architects and Marine Engineers (SNAME) Transactions 73:135-177.

Eda, H. and D. Savitsky. 1969. Experimental and Analytical Studies of Ship Controllability in Canals. Davidson Laboratory Technical Note 809. Hoboken, N.J.: Davidson Laboratory.

Eda, H., P. K. Shizume, J. S. Case, and J. J. Puglisi. 1986. A study of shiphandling performance in restricted water: Development and validation of computer simulation model. Society of Naval Architects and Marine Engineers (SNAME) Transactions 94:93-119.

Edwards, R. Y., Jr. 1985. Hydrodynamic forces on vessels stationed in a current. Pp. 99-106 in Proceedings of Seventeenth Annual Offshore Technology Conference (OTC), Houston, Texas, vol. 4. Paper OTC 5032. Richardson, Texas: OTC.

Elzinga, T. 1982. The Use of Ship Handling Simulators in Port Development and Operation. Paper presented at the International Ports Technology Conference (PORTECH 82), Singapore, Malaysia, June 22-26, 1982.

Eskola, H. 1986. Modelling the propulsion machinery behaviour during model propulsion tests in ice. In Proceedings of Polartech '86. Espoo, Finland: Technical Research Centre of Finland.

Finlayson, I. 1991. Creating a scene: Despite the computer, convincing simulator visuals depend on much artifice. Aerospace America (July):30-32.

Forsman, B., and J. Sandkvist. 1986. Brash ice effects on ship operations—A presentation of the SSPA manoeuvreing simulation model and other brash ice related projects. In Proceedings of Polartech '86. Espoo, Finland: Technical Research Centre of Finland.

Francingues, N. R., Jr. 1988. Technical management strategy for the disposal of dredged material containment testing and controls. Pp. 126-136 in Proceedings of International Seminar on Environmental Impact Assessment of Port Development, Baltimore, Maryland, November 12-19, 1988. London, England: International Maritime Organization.

Frankel, E. G. 1989. Strategic planning applied to shipping and ports. Maritime Policy Management 16(2):123-132.

Frisch, B. 1991. Simulating a new species of bird: Osprey must do more, and so must its simulator. Aerospace America (July):22-24.

Froese, J. 1988. Harbour and waterway planning by means of ship simulators. Pp. 1.21-1.35 in Proceedings of the Ninth International Harbour Congress Conference. Antwerp, Belgium: Royal Society of Flemish Engineers.

Fujii, H. 1972. On manoeuvre tests to investigate the course-keeping qualities of ships. Journal of Mechanical Engineering Science 14(7)(suppl.):70-74.

Fujino, M. 1968. Experimental studies on ship manoeuverability in restricted waters, Part 1. International Shipbuilding Progress 15(168):279-301.

Fujino, M. 1970. Experimental studies on ship manoeuverability in restricted waters, Part 2. International Shipbuilding Progress 17(186):45-65.

Fujino, M. 1982. Interim Report of the Activities of JAMP Working Group. Report to the International Towing Tank Conference (ITTC) Maneuvering Committee. October. Tokyo, Japan: Tokyo University.

Fujino, M., and T. Ishiguro. 1984. A study of the mathematical model describing manoeuvreing in shallow water—shallow water effects on rudder effectiveness parameters. Journal of Society of Naval Architects of Japan 156:180-192.

Gates, E. T. 1989. Maritime Accidents: What Went Wrong? Houston, Texas: Gulf Publishing Company.

Gates, E. T., and J. B. Herbich. 1978. A Mathematical Model for the Review of Deep-Draft Navigation Channels with Respect to Vessel Drift and Rudder Angles. Paper 12 in Proceedings of Symposium on Aspects of Navigability of Constraint Waterways, Including Harbor Entrances, vol. 2. Delft, Netherlands: Delft Hydraulics Laboratory.

Gertler, M., and G. R. Hagen. 1967. Standard Equations of Motion for Submarine Simulation. David Taylor Naval Ship Research and Development Center (NSRDC) Report 2510. June. Bethesda, Md.: NSRDC.

Gill, A. D. 1979. Mathematical Modelling of Ship Manoeuvreing. National Maritime Institute (NMI) Report R66. Feltham, England: NMI.

Glansdorp, C. C. 1975. Ship type modelling for a training simulator. Pp. 4-117–4-136 in

Proceedings of the Fourth Ship Control Systems Symposium, The Hague, Netherlands. Den Helder, Netherlands: Royal Netherlands Naval College.

Göransson, S., and H. Liljenberg. 1975. Simulating the Main Engine—A Comparison of FPP and CPP Arrangements. Swedish State Shipbuilding Tank (SSPA) Allmän Rapport 53. Gothenburg, Sweden: SSPA Maritime Consulting. 99 pages.

Grabowski, M. 1989. Decision aiding technology and integrated bridge design. Paper presented at Spring Meeting/STAR Symposium, Jersey City, N.J., April 12-15, 1989. Jersey City, N.J.: The Society of Naval Architects and Marine Engineers.

Graham, D. M. 1990. Markets proliferating for global positioning systems. Sea Technology (March):55-56.

Gress, R. K., and J. D. French. 1980. Systematic Evaluation of Navigation in Harbors. Pp. 568-583 in Ports '80. New York: American Society of Civil Engineers.

Gronarz, A. 1988. Maneuvering Behavior of a Full Ship in Water of Limited Depth. Versuchsaustalt für Binnenschittbau Duisburg (VBD) Report 1226. August. Duisburg, Germany: VBD.

Hagen, G. R. 1983. A catalog of existing mathematical models for maneuvering. Pp. 673-735 in D. Savitsky, J. F. Dalzel, and M. Palazzo, eds., Proceedings of Twentieth America Towing Tank Conference, August 2-4, 1983, vol. 2. Hoboken, N.J.: Stevens Institute of Technology.

Hanley, A. Simulators will help pick the LH winner: Simulators will approach the experience without the cost. Aerospace America (November):27-29.

Harlow, E. H. 1992. Harbor Planning and Design. In J.B. Herbich, ed., Handbook of Coastal Ocean Engineering, vol. 3: Harbors, Navigational Channels, Estuaries, and Environmental Effects. Houston, Texas: Gulf Publishing Company.

Heine, I. M. 1980. The U.S. Maritime Industry In the National Interest. National Maritime Council. Washington, D.C.: Acropolis Books.

Herbich, J.B. 1986. National Port Issues. Pp. 1031-1046 in P. H. Sorensen, ed., Ports '86. New York: American Society of Civil Engineers.

Herbich, J.B., ed. 1992. Handbook of Coastal Ocean Engineering, vol. 3: Harbors, Navigational Channels, Estuaries, and Environmental Effects. Houston, Texas: Gulf Publishing Company.

Hershman, M. J., ed. 1988. Urban Ports and Harbor Management. New York: Taylor & Francis.

Hess, R. A., and B. L. Hildreth. 1990. Numerical Identification and Estimation: An Efficient Method for Improving Simulator Fidelity. Paper presented at the Twelfth Interservice/Industry Training Systems Conference, November 5-8, 1990, Orlando, Florida.

Hewlett, J. C., and R. H. Nguyen. 1987. Applications of Ship Simulators to Ship Channel Design. Pp. 42-61 in Proceedings of San Francisco District Navigation Workshop, Sausalito and San Francisco, May 19-21, 1987. San Francisco, Calif.: U.S. Army Corps of Engineers District, San Francisco.

Hirano, M. 1980. On calculation method of ship manoeuvreing motion at initial design stage. Journal of Society of Naval Architects of Japan 147:144-153.

Hirano, M., J. Takashina, M. Fukushima, and S. Moriya. 1985. A Computer Program System for Ship Manoeuvreing Motion Prediction. Technical Bulletin 85-01. Tokyo, Japan: Mitsui Engineering and Shipbuilding.

Hochstein, A. B., R. Guido, and I. Zabaloieff. Optimization of channel maintenance expenses. Journal of Waterway Port Coast and Ocean Engineering 109(3):310-322.

Hoffman, D. 1972. Consideration of sea-keeping in the design of a ship manoeuvreing simulator. Paper 9 B-2, in Proceedings of the Third Ship Control Systems Symposium, Bath, England. Bath, England: Ministry of Defence.

Holzhüter, T. 1990. A workable dynamic model for the track control of ships. Pp. 4.275-

4.298 in Proceedings of the Ninth Ship Control Systems Symposium, Bethesda, Md., vol. 4. Washington, D.C.: Commander Naval Sea Systems Command.

Hooft, J. P. 1968. The manoeuverability of ships on a straight course. International Shipbuilding Progress 15(162):44-68.

Hooft, J. P. 1986. Computer simulations of the behavior of maritime structures. Marine Technology 23(2):139-156.

Hooyer, H. H. 1983. Behavior and Handling of Ships. Cambridge, Md.: Cornell Maritime Press.

Hu, P. N. 1961. Forward Speed Effect on Lateral Stability Derivatives of a Ship. Davidson Laboratory Report R-829. Hoboken, N.J.: Davidson Laboratory.

Huval, C., B. Comes, and R. T. Garnen III. 1985. Ship Simulation Study of John F. Baldwin (Phase 2) Navigation Channel San Francisco Bay, California. Department of the Army Technical Report HL-85-4, Hydraulics Laboratory Department. Vicksburg Miss.: U.S. Army Corps of Engineers Waterways Experiment Station.

Hwang, W-Y. 1985. The validation of a navigator model for use in computer aided channel design. Pp. A5-1—A5-18 in Proceedings of the Sixth Computer Aided Operations Research Facility (CAORF) Symposium. Kings Point, N.Y.: National Maritime Research Center.

Imlay, F. H. 1961. The Complete Expressions for Added Mass of a Rigid Body Moving in an Ideal Fluid. David Taylor Model Basin (DTMB) Report 1528. Bethesda, Md.: David Taylor Naval Ship Research and Development Center.

Inoue, S., and K. Murayama. 1970. Calculation of turning ship derivatives in shallow water. Higashi-ku Fukuoka-shi, Japan: West Japan Society of Naval Architects.

Inoue, S., M. Hirano, K. Kijima, and J. Takashima. 1981. A practical calculation method of ship manoeuvreing motion. International Shipbuilding Progress 28(325):207-222.

Jacobs, W. R. 1964. Estimation of Stability Derivatives and Indices of Various Ship Forms, and Comparison with Experimental Results. Davidson Laboratory Report 1035. Hoboken, N.J.: Davidson Laboratory.

Jensen, P. A., and J. M. Kieslich. 1986. Additional benefits from channel improvements. Pp. 920-931 in P. H. Sorensen, ed., Ports '86. New York: American Society of Civil Engineers.

Journal of Commerce. 1991a. Dredging battle but port is optimistic. April 30. Pp. 1C, 3C.

Journal of Commerce. 1991b. U.S. ports' conference to focus on search for money, recognition. March 20.

Kagan, R. A. 1990. The Dredging Dilemma: How Not to Balance Economic Development and Environmental Protection. Working Paper 90-3, Institute of Government Studies. Berkeley, Calif.: University of California at Berkeley.

Källström, C. G. 1979. Identification and Adaptive Control Applied to Ship Steering. Report LUTFD 2/(TFRT-1018/1-192. Department of Automatic Control. Lund, Sweden: Lund Institute of Technology.

Källström. C. G. 1984. A digital control system for ship manoeuvreing in ports and waterways. Pp. 163-179 in Proceedings of Seventh Ship Control Systems Symposium, vol. 3. Bath, England: Ministry of Defence.

Källström, C. G., and P. Ottosson. 1982. The generation and control of roll motion of ships in close turns. Pp. 22-36 in Proceedings of the Fourth International Symposium on Ship Operation Automation, Genova, Italy. New York: North-Holland Publishing Company.

Kaufman, E. J. 1985. Optimizing the use of compressed time simulation as a screening device for alternative channel layouts. Pp. C1-1–C1-8 in Proceedings, Sixth Computer Aided Operations Research Facility (CAORF) Symposium, Harbor and Waterway Development. Kings Point, N.Y.: National Maritime Research Center.

Kaufman, E.J. 1986. The use of simulation techniques for the development and validation of a proposed widening solution for the Gaillard Cut section of the Panama Canal. Paper presented at Nineteenth Dredging Seminar, Western Dredging Association Meeting, Baltimore, Md., October 15-17, 1986.

Khattab, O. 1987. Ship handling in harbours using real time simulation. Paper No. 11 in Proceedings of International Conference on Ship Manoeuvrability: Prediction and Achievement, April 29-30 and May 1, 1987. London, England: Royal Institute of Naval Architects.

Kijima, E., Y. Nakiri, Y. Tsutsui, and M. Matsunaga. 1990. Prediction method of ship maneuverability in deep and shallow waters. Pp. 311-318 in Proceedings of the Joint International Conference on Marine Simulation (MARSIM 90) and Ship Maneuverability (ICSM '90). Tokyo, Japan: Society of Naval Architects of Japan.

Kirchhoff, G. 1869. Über die Bewegung eines Rotationskörpers in einer Flüssigkeit. Crelle 71(237).

Kobayashi, E. 1988. A Simulation Study on Ship Manoeuverability at Low Speeds. Technical Bulletin no. 180. Nagasaki, Japan: Mitsubishi Heavy Industries.

Koelink, J. Th. H. 1968. Approximate methods in Z-steering test analysis. International Shipbuilding Progress 15(162):35-43.

Kose, K. 1982. On a new mathematical model of manoeuvreing motions of a ship and its applications. International Shipbuilding Progress 29(336):205-220.

Kotschin, N. J., I. A. Kibel, and N. W. Rose. 1954. Theoretische Hydromechanik. Bund I (from Russian). Berlin: Akademie-Verlag.

Koyama, T. 1972. Improvement of course stability and control by a subsidiary automatic control. Journal of Mechanical Engineering Science 14(7)(suppl.):132-141.

Koyama, T., and Y. Jin. 1987. An expert system approach to collision avoidance. Pp. 234-263 in Proceedings of the Eighth Ship Control Systems Symposium, vol. 13. The Hague, Netherlands: Ministry of Defense, Netherlands.

Koyama, T., K. Kose, and K. Hasegawa. 1977. A study of the instability criterion on the manual steering of ships. Journal of the Society of Naval Architects of Japan 142 (December):119-126.

Kristiansen, S., E. Rensvik, and L. Mathisen. 1989. Integrated total control of the bridge. Society of Naval Architects and Marine Engineers (SNAME) Transactions 97:279-297.

Lamb, H. 1918. The Inertia Coefficients of an Ellipsoid Moving in a Fluid. ARC R & M No. 623. London, England: Aeronautical Research Council.

Landweber, L., and M. C. de Macagno. 1957. Added mass of two-dimensional forms oscillating in a free surface. Journal of Ship Research 1(3):20-30.

Lindström, C. A. 1989. Prediction of Ship Manoeuvreing in Level Ice by Simulation of the Planar Motions (in Swedish). Memorandum M-7. Otaniemi, Finland: Arctic Offshore Research Center.

Loman, G. J. A., and R. J. M. van Maastrigt. 1988. Manoeuvreing Risks in Confined Fairways, Assessment Through Simulation. Pp. 5.199-5.211 in Proceedings of the Ninth International Harbour Conference. Antwerp, Belgium: Royal Society of Flemish Engineers.

Mandel, P. 1967. Ship Maneuvering and Control. Pp. 463-606 in J. P. Comstock, ed., Principles of Naval Architecture. New York: Society of Naval Architects and Marine Engineers.

MacElrevey, D. H. 1988. Shiphandling for the Mariner. Centerville, Md.: Cornell Maritime Press.

Maconachie, B. 1990. The navigation union: Some integrated bridge systems reviewed. Naval Architect April:E175, E177-E178, E179.

Marine Board. 1985. Background Paper of the Technical Panel on Ports, Harbors, and Navigation Channels of the Committee on National Dredging Issues. Unpublished background paper, Marine Board, National Research Council, Washington, D.C.

Maritime Research Institute Netherlands. 1988. Fast or real-time simulation? Marin Report 33 (September):378-379

Matsumoto, N. and K. Suemitsu. 1984. Interference effects between hull, propeller and rudder on a hydrodynamic mathematical model in maneuvering motion. Naval Architecture and Ocean Engineering 22:14-126.

Matthews, R. 1984. A six degree of freedom ship model for computer simulation. Pp. 365-374 in Proceedings of the Third International Conference on Marine Simulation (MARSIM 84). Rotterdam, Netherlands: Maritime Research Institute Netherlands.

McAleer, J. B., C. F. Wicker, and J. R. Johnson. 1965. Design of Channels for Navigation. Chapter 10 in Report No. 3, Committee on Tidal Hydraulics. Vicksburg, Miss.: U.S. Army Corps of Engineers, Waterways Experiment Station.

McCallum, I. R. 1976. A New Approach to Manoeuvreing Ship Simulation. Ph.D. dissertation. London: City University, London.

McCallum, I. R. 1980. A ship steering mathematical model for all manoeuvreing regimes. Pp. 119-141 in Proceedings of Symposium on Ship Steering Automatic Control, June 25-27, 1980, Genova, Italy. Genova, Italy: Internazionale Delle Comunicazioni.

McCallum, I. R. 1982. New Simulation Techniques in Harbour Design. Paper presented at 'PORTECH 82,' International Ports Technology Conference, Singapore, Malaysia, June 22-26, 1982.

McCallum, I. R. 1984. Which model? A critical survey of ship simulator mathematical models. Pp. 375-388 in Proceedings of the Third International Conference of Marine Simulation (MARSIM 84). Rotterdam, Netherlands: Maritime Research Institute Netherlands.

McCallum, I. R. 1987. Dynamic Quantitative Methods for Solving Actual Ship Manoeuverability Problems. Paper presented at Royal Institute of Naval Architects (RINA) Conference on Ship Manoeuvrability: Prediction and Achievement, April 29-May 1, 1987, London. Mid Glamorgan, Wales: Maritime Dynamics.

McCartney, B. L. 1985. Deep Draft Navigation Project Design. Journal of Waterway, Port, Coastal and Ocean Engineering 3(1):18-28.

McCreight, W. R. 1986. Ship maneuvering in waves. Pp. 456-469 in Proceedings of the Sixteenth Office of Naval Research (ONR) Symposium on Naval Hydrodynamics, Berkeley, Calif. Berkeley, Calif.: Department of Naval Architecture and Offshore Engineering, University of California, Berkeley.

McEwen, W. A. and A. H. Lewis. 1953. Encyclopedia of Nautical Knowledge. Cambridge, Md.: Cornell Maritime Press.

McIlroy, W., H. Grossman, H. Eda, and P. Shizume. 1981. Validation Procedures for Ship Motion and Human Perception. Paper presented at the Second International Conference on Marine Simulation Symposium (MARSIM 81), June 1-5, 1981, National Maritime Research Center, Kings Point, N.Y.

Mikelis, N. E., A. J. P. S. Clarke, S. J. Roberts, and E. H. A. J. Jackson. 1985. On the construction of a versatile mathematical model for marine simulation. Pp. 75-88 in M. Heller, ed., Proceedings of the First Intercontinental Marine Simulation Symposium, Munich, Germany. New York: Springer-Verlag.

Miller, E. R., Jr. 1979. Towboat Maneuvering Simulator. Vol. 3, Theoretical Description. U.S. Coast Guard Report CG-D63-79. Washington, D.C.: U.S. Coast Guard.

Miller, E. R., Jr. 1980. Model Test and Simulation Correlation Study Based on the Full Scale *Esso Osaka* Maneuvering Data. Technical Report 8007-1. Laurel, Md.: Hydronautics.

Moody, C.G. 1970. Study of the Performance of Large Bulk Cargo Ships in a Proposed Interoceanic Canal. Report Number NSRDC/HMD-0374-H-01 (proprietary report - limited distribution). Bethesda, Md.: David Taylor Naval Ship Research and Development Center.

Moraal, J. 1980. Evaluating simulator validity. Pp. 411-426 in Manned Systems Design: Methods, Equipment, and Applications. New York: Plenum Press.

Motora, S. 1960. On the Measurement of Added Mass and Added Moment of Inertia of Ships in Steering Motion. Pp. 241-274 in Proceedings of the First Symposium on Ship Maneuverability. David Taylor Model Basin Report 1461. Washington, D.C.: David Taylor Naval Ship Research and Development Center.

Motora, S., M. Fujino, M. Sugiura, and M. Sugita. 1971. Equivalent Added Mass of Ships in the Collision. Selected Papers of the Society of Naval Architects of Japan 7:138-148.

Naegle, J. N. 1980. Ice Resistance Prediction and Motion Simulation for Ships Operating in the Continuous Mode of Icebreaking. Ph.D. dissertation. University of Michigan, Ann Arbor.

National Research Council. 1981. Problems and Opportunities in the Design of Entrances to Ports and Harbors. Washington, D.C.: National Academy Press.

National Research Council. 1983. Criteria for the Depths of Dredged Navigation Channels. Washington, D.C.: National Academy Press.

National Research Council. 1985. Dredging Coastal Ports: An Assessment of the Issues. Washington, D.C.: National Academy Press.

National Research Council. 1986. Advances in Environmental Information Services for Ports: An Assessment of Uses and Technology. Washington, D.C.: National Academy Press.

National Research Council. 1987. Sedimentation Control to Reduce Maintenance Dredging for Navigational Facilities in Estuaries. Washington, D.C.: National Academy Press.

Newman, J. N. 1966. Some hydrodynamic aspects of ship maneuverability. Pp. 33-58 in Proceedings of the Sixth Symposium on Naval Hydrodynamics. Washington, D.C.: Government Printing Office.

Newman, J. N. 1969. Lateral motion of a slender body between two parallel walls. Journal of Fluid Mechanics 39(Part 1):97-115.

Newman, J. N. 1972. Some Theories for Ship Manoeuvreing. Paper prepared for Massachusetts Institute of Technology, Contract No. N00014-67-A-0240023. Springfield, Va.: National Technical Information Service.

Nikolaev, E. P., R. Y. Pershits, and A. A. Russetzky. 1972. Estimation of the effectiveness of lateral thrust units. Journal of Mechanical Engineering Science 14(7)(suppl.):155-160.

Nomoto, K. 1960. Analysis of Kempf's Standard Manoeuvre Test and Proposed Steering Quality Indices. Pp. 275-304 in Proceedings of the First Symposium on Ship Maneuverability. David Taylor Model Basin Report 1461. Washington, D.C.: David Taylor Naval Ship Research and Development Center.

Nomoto, K. 1966. Unusual Scale Effects on Manoeuverabilities of Ships with Blunt Bodies. Pp. 554-556 in Proceedings of the Eleventh International Towing Tank Conference (ITTC). Tokyo, Japan: The Society of Naval Architects of Japan.

Nomoto, K. 1972. Problems and requirements of directional stability and control of surface ships. Journal of Mechanical Engineering Science 14(7)(suppl.):1-5.

Nomoto, K., T. Taguchi, K. Honda, and S. Hirano. 1957. On the steering qualities of ships. International Shipbuilding Progress 4(35):354-370.

Norrbin, N. H. 1960. A Study of Course Keeping and Manoeuvreing Performance. Swedish State Shipbuilding Tank (SSPA) Publication No. 45. Gothenburg, Sweden: SSPA Maritime Consulting. Also Pp. 359-423 in Proceedings of the First Symposium on Ship Maneuverability. David Taylor Model Basin Report (DTMB) 1461. Bethesda, Md.: David Taylor Naval Ship Research and Development Center.

Norrbin, N. H. 1963. On the design and analysis of the zig zag test on base of quasi-linear frequency response. Pp. 355-374 in Proceedings of the Tenth International Towing Tank Conference, London, England. Teddington, England: National Physics Laboratory.

Norrbin, N. H. 1965. The Technique and Analysis of the Zig Zag Test (in Swedish). Swedish

State Shipbuilding Tank (SSPA) Allmän Rapport No. 12. Gothenburg, Sweden: SSPA Maritime Consulting.

Norrbin, N. H. 1970. Theory and observations on the use of a mathematical model for ship manoeuvreing in deep and confined waters. Pp. 807-904 in Proceedings of the Eighth Symposium on Naval Hydrodynamics, Pasadena, Calif. Arlington, Va.: Office of Naval Research.

Norrbin, N. H. 1972. Ship manoeuvreing with application to shipborne predictors and real-time simulators. Journal of Mechanical Engineering Science 14(7)(suppl.):91-107.

Norrbin, N. H. 1974. Bank effects on a ship moving through a short dredged channel. Pp. 71-88 in Proceedings of the Tenth Symposium on Naval Hydrodynamics, Cambridge, Mass. Arlington, Va.: Office of Naval Research.

Norrbin, N. H. 1978. A method for the prediction of the manoeuvreing lane of a ship in a channel of varying width. Pp. 22.1-22.16 in Proceedings of the Symposium on Aspects of Navigability of Constrained Waters, Including Harbour Entrances, vol. 3. Delft, Netherlands: Delft Hydraulics Laboratory.

Norrbin, N. H. 1986. Fairway Design with Respect to Ship Dynamics and Operational Requirements. Swedish State Shipbuilding Tank (SSPA) Research Report No. 102. Gothenburg, Sweden: SSPA Maritime Consulting.

Norrbin, N. H. 1988. Head-on collisions or a planned encounter—A contribution to micro-navigation. In Proceedings of the Sixth International Symposium on Vessel Traffic Services, Gothenburg, Sweden. Gothenburg, Sweden: Port of Gothenburg Authority.

Norrbin, N. H. 1989. Ship Turning Circles and Manoeuvreing Criteria. Pp. S-6-2-12–S-6-2-12 in Proceedings of the Fourteenth Ship Technology and Research (STAR) Symposium, 1989, No. SY-25. Jersey City, N.J.: Society of Naval Architects and Marine Engineers.

Norrbin, N. H., S. Göransson, R. J. Risberg, and D. H. George. 1978. A study of the safety of two-way traffic in a Panama Canal bend. Pp. K1 3.1-3.36 in Proceedings of the Fifth Ship Control Systems Symposium, Annapolis, Md., vol. 3. Bethesda, Md.: David Taylor Naval Ship Research and Development Center.

Ogawa, A., and H. Kasai. 1978. On the mathematical model of manoeuvreing motion of ships. International Shipbuilding Progress 25(292):306-319.

Olson, H. E., J. R. Hanchey, L. G. Antle, A. F. Hawn, D. F. Bastian, and D. V. Grier. 1986. Planning for deep draft navigation channels. Pp. 784-796 in P. H. Sorensen, ed., Ports '86. New York: American Society of Civil Engineers.

Oltmann, P., and S. D. Sharma. 1985. Simulation of combined engine and rudder manoeuvres using an improved model of hull-propeller-rudder interactions. Pp. 83-108 in Proceedings of the Fifteenth Office of Naval Research Symposium on Naval Hydrodynamics, Hamburg, Germany. Washington, D.C.: National Academy Press.

Onassis, I., and D. ten Hove. 1988. Modular Ship Manoeuvreing Models. Netherlands Organization for Applied Scientific Research (TNO) Report 5163034-88-1. Delft, The Netherlands: TNO.

Ottosson, P., and L. Apleberger. 1987. Real-time simulations of ship motions—A tool for the design of a new pacific port. In Nanjing Hydraulic Research Institute, ed., Proceedings, Second International Conference on Coastal and Port Engineering in Developing Countries, Beijing, Peoples Republic of China. Nanjing, China: China Ocean Press.

Ottosson, P., and W. B. van Berlekom. 1985. Simulations in real and accelerated time—A computer study of a Ro-Ro vessel entering different port configurations. In Proceedings of Association Technique et Maritime (ATMA) Spring Session. Paris, France: ATMA.

Paffett, J. A. H. 1981. Recent developments in marine simulation. The Journal of Navigation 34(2):165-186.

Paloubis, J., and J. G. Thaler. 1972. Identification of System Models from Operating Data.

Paper IB-3 in Proceedings of the Third Ship Control Systems Symposium. Bath, England: Ministry of Defence.

Perdok, J., and T. H. Elzinga. 1984. The application of micro-simulators in port design and shiphandling training courses. Pp. 215-226 in Proceedings of the Third International Conference on Marine Simulation (MARSIM 84). Rotterdam, Netherlands: Maritime Research Institute Netherlands.

Perdok, J., and C. van der Tak. 1987. The application of man-machine models in the analysis of ship control. Pp. 228-237 in Proceedings of the Eighth Ship Control Systems Symposium, vol. 13. The Hague, Netherlands: Ministry of Defense, Netherlands.

Perez y Perez, L. P. 1972. A Time Domain Solution to the Motions of a Steered Ship in Waves. Prepared by Department of Naval Architecture, University of California, Berkeley, for U.S. Coast Guard (USCG). Report CG-D-19-73. Washington, D.C.: USCG.

Permanent International Association of Navigation Congresses (PIANC). 1980. Optimal Layout and Dimensions for the Adjustment to Large Ships or Maritime Fairways in Shallow Seas, Seastraits and Maritime Waterways. International Commission for the Reception of Large Ships, PIANC Bulletin No. 35(suppl.). Brussels, Belgium: PIANC.

Permanent International Association of Navigation Congresses (PIANC). 1985. Underkeel Clearance of Large Ships in Maritime Fairways with Hard Bottom. Report of Working Group of P.T.C. 2. Brussels, Belgium: PIANC.

Plummer, Carlyle J. 1966. Ship Handling in Narrow Channels (second edition). Cambridge, Md.: Cornell Maritime Press.

Pourzanjani, M. 1990. Formulation of the force mathematical model of ship manoeuvreing. International Shipbuilding Progress 37(409):5-32.

Pourzanjani, M. M., I. R. McCallum, J. O. Flower, and H. K. Zienkiewicz. 1987. Hydrodynamic lift and drag simulation for ship manoeuvreing models. Pp. 340-349 in Proceedings of the Fourth International Conference on Marine Simulation (MARSIM '87), Trondheim, Norway, June 22-24, 1987. Trondheim, Norway: International Marine Simulator Forum.

Price, W. G. 1972. The stability of a ship in a simple sinusoidal wave. Journal of Mechanical Engineering Science 14(7)(suppl.):194-206.

Puglisi, J. J. 1985. CAORF's cooperative role in the application of simulation to optimizing channel design and maintenance. Pp. A2-1–A2-20 in Proceedings of the Sixth Computer Aided Operations Research Facility (CAORF) Symposium: Harbor and Waterway Development. Kings Point, N.Y.: National Maritime Research Center.

Puglisi, J. J. 1986. The requirements and application for the use of simulation techniques at CAORF as an engineering design tool in the ship navigation, channel design and maintenance optimization process. Pp. 59-95 in Proceedings of the Nineteenth Dredging Seminar, Baltimore, Md., October 15-17, 1986. College Station, Tex.: Texas A&M Sea Grant College Program.

Puglisi, J. J. 1987. History and Future Development in the Application of Marine Simulators, Tomorrow's Challenging Role for the International Marine Simulator Forum (IMF). Pp. 5-29 in Proceedings of the Fourth International Conference on Marine Simulation (MARSIM '87), Trondheim, Norway, June 22-24, 1987. Trondheim, Norway: International Marine Simulator Forum.

Puglisi, J. J. 1988. Use of Simulation Techniques, Capabilities and Methodology Required for Harbor/Waterway Design Studies. In T. K. S. Murphy, ed., Addendum to Proceeding/Tutorial Sessions, Second International Conference on Computer Aided Design Manufacture and Operation in the Marine and Offshore Industries, Southampton, England. Boston, Mass.: Computational Mechanics Publications.

Puglisi, J. J., J. S. Case, and W-Y Hwang. 1985a. Direction of international joint effort for development of mathematical models and ship performance data for marine simulation application. Pp. 88-101 in M. R. Heller, ed., Proceedings of the First Intercontinental Marine Simulation Symposium, Munich, Germany. Munich: Springer-Verlag.

Puglisi, J. J., J. S. Case, and W-Y Hwang. 1985b. The progress of validation in simulation research. Pp. A1-1–A1-12 in Proceedings of the Sixth Computer Aided Operations Research Facility (CAORF) Symposium. Kings Point, N.Y.: National Maritime Research Center.

Puglisi, J., and A. D'Amico. 1985. Fidelity of Simulation in Harbor and Waterway Development. Pp. F2-1–F2-4 in Proceedings of the Sixth Computer Aided Operations Research Facility (CAORF) Symposium: Harbor and Waterway Development. Kings Point, N.Y.: National Maritime Research Center.

Puglisi, J., A. D'Amico, and G. Van Hoorde, G. 1984. The Use of Simulation at CAORF in Determining Criteria for Increased Throughput of Ship Traffic in the Panama Canal. Pp. 239-248 in Proceedings of the Third International Conference of Marine Simulation (MARSIM 84). Rotterdam, Netherlands: Maritime Research Institute Netherlands.

Puglisi, J., G. van Hoorde, E. Kaufman, and H. Eda. 1987. The Proposed Plan for Widening of the Panama Canal and Application of Simulator Techniques for the Development and Validation of the Proposed Solution. Pp. 166-192 in Proceedings of the Fourth International Conference on Marine Simulation (MARSIM '87), June 22-24, 1987. Trondheim, Norway: International Marine Simulator Forum.

Reid, G. H. 1975. Primer of Towing. Cambridge, Md.: Cornell Maritime Press.

Reid, G. H. 1986. Shiphandling with Tugs. Centerville, Md.: Cornell Maritime Press.

Renilsson, M. R., and A. Driscoll. 1982. Broaching—An investigation into the loss of directional control in severe following seas. Transactions of the Royal Institute of Naval Architects (RINA) 124:253-273.

Report of the Maneuvering Committee. 1987. Pp. 345-400 in Proceedings of the Eighteenth International Towing Tank Conference, vol. 1. Tokyo, Japan: Society of Naval Architects of Japan.

Rogers, J. G. 1984. Origins of Sea Terms. Boston, Mass.: Nimrod Press.

Rosselli, A., C. J. Sobremisana, and G. Muller. 1990. Changing directions for traditional ports. Pp. 113-119 in Proceedings of the Twenty-Seventh International Navigation Congress, Section 2, Maritime Ports and Seaways (for commercial, fishery, and pleasure navigation), May. Brussels, Belgium: Permanent International Association of Navigation Congresses.

Russell, P. J. D. 1987. A mariner's view of the user requirements of the electronic chart and electronic chart display and information systems. Seaways (December):3-5.

Rutgersson, O., and P. Ottosson. 1987. Model tests and computer simulations—An effective combination for investigation of broaching phenomena. Society of Naval Architects and Marine Engineers (SNAME) Transactions 95:263-281.

Rydill, L. J. 1959. A linear theory for the steered motion of ships in waves. Transactions of the Royal Institute of Naval Architects (RINA) 101:81-112.

Salo, M. and M. Heikkilä. 1990. On the modelling of hull-propeller-rudder interactions in manoeuvreing of twin-screw ships. Supplemental paper in Proceedings of the Joint International Conference on Marine Simulation (MARSIM 90) and Ship Maneuverability (ICSM '90), Tokyo, Japan: Society of Naval Architects of Japan.

Sandvik, R. 1990. Updating the electronic chart: The SEATRANS project. International Hydrographic Review 67(July):59-67.

Sargent, T. P., and P. Kaplan. 1970. System Identification of Surface Ship Dynamics. Technical Report 70-72. Plainview, N.Y.: Oceanics.

Schmidt, M., and K. H. Unterreiner. 1976. Ein mathematisches modell zur simulation des manövrierverhaltens von schiffen für die anwendung in trainings-simulatoren. Shiff und Hafen 11(11).

Schoenherr, K. E. 1960. Data for estimating bank suction effects in restricted water and on merchant ship hulls. Pp. 199-210 in Proceedings of the First Symposium on Ship Maneuverability. David Taylor Model Basin Report 1461. Bethesda, Md.: David Taylor Naval Ship Research and Development Center.

Schuffel, H. 1984. On the interaction of rate of movement presentation with ship's controllability. Pp. 151-159 in Proceedings of the Third International Conference on Marine Simulation (MARSIM 84). Rotterdam, Netherlands: Maritime Research Institute Netherlands.

Seymour, R. J., and J. R. Vadus. 1986. Real time information systems for improved port operations. Pp. 515-526 in P. H. Sorensen, ed., Ports '86. New York: American Society of Civil Engineers.

Shooman, M. L. 1980. Models of Helmsman and Pilot Behavior for Manoeuvreing Ships. Computer Aided Operations Research Facility (CAORF) Technical Memorandum, CAORF 40-7901-02. Kings Point, N.Y.: National Maritime Research Center.

Simoen, R., P. Kerckaert, D. Vandenbossche, and L. Neyrinck. 1980. Extension of the Port of Zeebrugge. Permanent International Association of Navigation Congresses (PIANC) Bulletin No. 37. Brussels, Belgium: PIANC.

Sjoberg, B. E. 1984. Special Features of the Planning and Construction of Channels in Finland. Permanent International Association of Navigation Congresses (PIANC) Bulletin No. 46. Brussels, Belgium: PIANC.

Smith, D. 1990. Flight simulation. Aerospace America (December):27.

Smitt, L. W. 1970. Steering and manoeuvreing: Full-scale and model tests. European Shipbuilding. Journal of the Ship Technical Society 19(6):86-96.

Smitt, L. W., and M. S. Chislett. 1972. Course stability while stopping. Journal of Mechanical Engineering Science 14(7)(suppl.):181-185.

Society of Naval Architects and Marine Engineers (SNAME). 1950. Nomenclature for treating the motion of a submerged body through a fluid. SNAME Technical and Research Bulletin No. 1-5. New York: SNAME.

Stix, G. 1991. Trends in transportation: Along for the ride? Scientific American 265(1):94-99, 102, 104-106.

Strøm-Tejsen, J. 1965. A Digital Computer Technique for Prediction of Standard Maneuvers of Surface Ships. David Taylor Model Basin Report No. 2130. Bethesda, Md.: David Taylor Naval Ship Research and Development Center.

Strøm-Tejsen, J., and M. S. Chislett. 1966. A model testing technique and method of analysis for the prediction of steering and manoeuvreing qualities of surface ships. Pp. 317-382 in Proceedings of the Sixth Symposium on Naval Hydrodynamics. Arlington, Va.: Office of Naval Research.

Stuurman, A. M. 1969. Modelling the helmsman: A study to define a mathematical model describing the behavior of a helmsman steering a ship along a straight course. Pp. 665-667 in Proceedings of the Twelfth International Towing Tank Conference. Rome, Italy: Instituto Natzionale per Studi ed Esperienze di Architettura Navale.

Tasai, F. 1961. Hydrodynamic force and moment produced by swaying oscillation of cylinders on the surface of a fluid. Journal of the Society of Naval Architects of Japan 110:9-17.

Thöm, H. 1975. Modellbildung für das Kursverhalten von Schiffen. Dissertation D17. Darmstadt, Germany: Techn Hochschule Darmstadt. 212 pages.

Trägårdh, P. 1976. Simulation of tugs at the SSPA manoeuvreing simulator. In Proceedings of the Fourth International Tug Convention, New Orleans, La., 1975. London, England: Thomas Reed Industrial Press.

Trankle, T. L., J. H. Vincent, and S. N. Franklin. 1981. Integrated Flight Testing Based on Nonlinear System Identification Data Processing Techniques. American Institute of Aeronautics and Astronautics (AIAA) Paper 81-2449 presented to AIAA First Flight Testing Conference, November 11-13, 1981, Las Vegas, Nevada.

Trankle, T. L., J. H. Vincent, and S. N. Franklin. n.d. System Identification of Nonlinear Aerodynamics Models. Advisory Group for Aerospace Research and Development (AGAR) Dograph No. 256, Advances in the Techniques and Technology of the Application of Nonlinear Filters and Kalman Filters. Neuilly Sur Seine, France: AGAR.

Tuck, E. O. 1966. Shallow water flows past slender bodies. Journal of Fluid Mechanics 26(Part 1):81-95.

Tuck, E. O., and J. N. Newman. 1974. Hydrodynamic interactions between ships. Pp. 28-51 in Proceedings of the Tenth Symposium on Naval Hydrodynamics, Boston, Mass. Arlington, Va.: Office of Naval Research.

U.S. Army Corps of Engineers. 1977. Shore Protection Manual, 3d ed. Vol. 1-3. Coastal Engineering Research Center. Washington, D.C.: U.S. Government Printing Office.

U.S. Army Corps of Engineers. 1983. Hydraulic Design of Deep Draft Navigation Projects. Engineer Manual EM 1110-2-1613. Washington, D.C.: U.S. Government Printing Office.

U.S. Army Corps of Engineers (USACE). 1986a. General Design Memo 1, Norfolk Harbor and Channels, Virginia, Channel Design and Simulation Studies, Appendix D, June. Norfolk, Va.: USACE District, Norfolk.

U.S. Army Corps of Engineers (USACE). 1986b. Report on Ship Simulator Capability and Channel Design. Report 86-R-2. Fort Belvoir, Va.: Institute for Water Resources, USACE.

U.S. Army Corps of Engineers (USACE). 1988. Oakland Harbor Deep Draft Navigation Improvements Design Memorandum No. 1, General Design, March. San Francisco, Calif.: USACE District, San Francisco.

U.S. Army Corps of Engineers (USACE). 1991. Savannah Harbor, Georgia, Comprehensive Study: Draft Feasibility Report and Environmental Impact Statement, April. Savannah, Ga.: USACE District, Savannah.

U.S. Coast Guard (USCG). 1987. Commandant Instruction 16500.11B: Waterway Analysis and Management System (WAMS). Washington, D.C.: USCG.

U.S. Department of Commerce. Statistical Abstract of the United States. Bureau of Census, various years. Washington, D.C.: U.S. Government Printing Office.

U.S. Maritime Administration. 1979. Validation of Computer Aided Operations Research Facility (CAORF). CAORF Technical Report, CAORF 90-7801-01. Kings Point, N.Y.: National Maritime Research Center. 80 pages.

U.S. Maritime Administration. 1980. Empirical Assessment of Comparative Shiphandling Performance as a Function of Vessel Class in the Port of Coatzacoalcos, Mexico. Computer Aided Operations Research Facility (CAORF) Report No. 24-8013-01, June. Kings Point, N.Y.: National Maritime Research Center.

U.S. Maritime Administration. 1981. Effect of Ship Size and Draft on Piloting Shiphandling Performance at Coatzacoalcos, Mexico. Computer Aided Operations Research Facility (CAORF) Report No. 24-8119-01. June. Kings Point, N.Y.: National Maritime Research Center.

U.S. Maritime Administration. 1983-1986. Computer Aided Operations Research Facility (CAORF) Final Reports prepared for the Panama Canal Improvement Project. Kings Point, N.Y.: CAORF.

U.S. Maritime Administration. 1985. The Application of CAORF as a New Technology for the Determination of Channel Navigability and Width Requirements, for: "Newport News Channel," 1983, and "Thimble Shoal Channel. Kings Point, N.Y.: Computer Aided Operations Research Facility.

U.S. Maritime Administration. 1986. Final Report-Validation of Proposed Widening Solution for the Gaillard Cut. Prepared for the Panama Commission, Canal Improvements Division, Balboa, Panama, September. Kings Point, N.Y.: Computer Aided Operations Research Facility.

U.S. Maritime Administration. 1987. Final Report, An Evaluation of Alternative Channel and Turning Basin Designs for the Inner and Outer Harbors of Oakland, Calif., June. Kings Point, N.Y.: Computer Aided Operations Research Facility.

U.S. Maritime Administration. 1990. A Report to the Congress on the Status of the Public Ports of the United States 1988-1989. April. Washington, D.C.: U.S. Department of Transportation.

U.S. Navy Hydrographic Office. 1956. Navigation Dictionary. H.O. Publication 220. Washington, D.C.: U.S. Government Printing Office.

Van Amerongen, J., and P. G. M. van der Klugt. 1985. Modelling and Simulation of the Roll Motions of a Ship. Pp. 161-175 in Proceedings of the First Intercontinental Maritime Symposium, Munich. New York: Springer-Verlag.

Van de Beek, H. 1987. Considerations on validation in navigation research and training. Pp. 303-309 in the Proceedings of Simulators 4, Orlando, Fla., April 6-9, 1987. San Diego, Calif.: Society for Computer Simulation.

Van Berlekom, W. B. 1978. Simulator investigations of predictor steering systems for ships. Transactions of the Royal Institute of Naval Architects (RINA) 120:23-34.

Van Berlekom, W. B., and T. A. Goddard. 1972. Maneuvering of large tankers. Society of Naval Architects and Marine Engineers (SNAME) Transactions 80:264-298.

Van Leeuwen, G. 1964. The Lateral Damping and Added Mass of an Oscillating Shipmodel. Publication No. 23. Delft, Netherlands: Shipbuilding Laboratory.

Van Leeuwen, G. 1972a. Some aspects of prediction and simulation of manoeuvres. Journal of Mechanical Engineering 14(7)(suppl.):108-114.

Van Leeuwen, G. 1972b. Course Keeping Going Astern. Paper III B-4 in Proceedings of the Third Ship Controls Systems Symposium. Bath, England: Ministry of Defence.

Van Maanen, W. Ph., ed. 1989. Guidelines on Maritime Pilot Training: User's Specifications, Working Group on Pilot Training. Toledo, Ohio: International Marine Simulator Forum.

Veldhuijzen, W. 1989. Vessel Traffic Engineering Research. Paper presented at the Netherlands Institute of Navigation Seminar, June 21, 1989, at Soesterberg. Delft, Netherlands: Netherlands Organization for Applied Scientific Research.

Veldhuijzen, W., and H. G. Stassen. 1975. Simulation of Ship Manoeuvreing Under Human Control. Pp. 6-148–6-163 in Proceedings of the Fourth Ship Control Systems Symposium, The Hague, Netherlands. Den Helder, Netherlands: Royal Netherlands Naval College.

Vugts, J. H. 1968. The Hydrodynamic Coefficients for Swaying, Heaving and Rolling Cylinders in a Free Surface. Report No. 194. Delft, Netherlands: Laboratorium voor Scheepsbouwkunde.

Walden, D. A., and R. K. Gress. 1981. A Survey of Coast Guard Simulator Development Research and Simulator Applications. Paper presented at the Second International Conference on Marine Simulation (MARSIM 81), June 1-5, 1981, Maritime Research Center, Kings Point, N.Y.

Waller, J. O., and D. A. Schuldt. n.d. Design Studies for Grays Harbor Navigation Improvement Project, Washington. Civil Projects Management Section. Seattle, Wash.: U.S. Army Corps of Engineers District, Seattle.

Webster, W. C. 1967. Analysis of the control of activated antiroll tanks. Society of Naval Architects and Marine Engineers (SNAME) Transactions 75:296-331.

Weinblum, G. P. 1952. On the directional stability of ships in calm water and in a regular seaway. Pp. 43-47 in Proceedings of the First U.S. National Congress of Applied Mechanics. New York: American Society of Mechanical Engineers.

Wendel, A. H., and J. F. Dunne. 1969. Dynamic analysis and simulation of ship and propulsion plant manoeuvreing performance. Pp. IV-A-1–IV-A-30 in Proceedings of the Second Ship Control Systems Symposium, Annapolis, Md. Annapolis, Md.: Naval Ship Research Development Laboratory.

Whalin, R. W. 1986. Preliminary Findings Report on Grays Harbor Ship Simulation Tests of Outer Harbor Scenario. Hydraulics Laboratory, December. Vicksburg, Miss.: U.S. Army Corps of Engineers Waterways Experiment Station.

Whalin, R. W. 1987. Preliminary Findings Report on Grays Harbor Ship Simulation Tests of Inner Harbor Region Scenario. Hydraulics Laboratory, March. Vicksburg, Miss.: U.S. Army Corps of Engineers Waterways Experiment Station.

Williams, K., J. Goldberg, and M. Gilder. 1982. Validation of the Effectiveness of Simulator Experience vs. Real-World Operational Experience in the Port of Valdez. Computer Aided Operations Research Facility (CAORF) Technical Report, CAORF 50-7917-02. Kings Point, N.Y.: National Maritime Research Center.

Wind, H. G., and M. J. Officier. 1984. Computation of shipmanoeuvring coefficients. International Shipbuilding Progress 31(36):277-284.

Yeung, R. W. 1978. Applications of slender body theory to ships moving in restricted shallow water. Paper No. 28 in Proceedings of the Symposium on Aspects of Navigability of Constrained Waterways, Including Harbour Entrances, vol. 3. Delft, Netherlands: Delft Hydraulics Laboratory.

Yeung, R. W. 1978. On the interactions of slender ships in shallow water. Journal of Fluid Mechanics 85:143-159.

Zhao, Y. 1990. Theoretical determination of ship manoeuvreing motion in shallow water. Pp. 493-503 in Proceedings of the Joint International Conference on Marine Simulation (MARSIM 90) and Ship Maneuverability (ICSM '90). Tokyo, Japan: Society of Naval Architects of Japan.

Zhou, W. W., and M. Blanke. 1987. Nonlinear recursive prediction error method applied to identification of ship steering dynamics. Pp. 85-106 in Proceedings of the Eighth Ship Control Systems Symposium, vol. 3. The Hague, Netherlands: Ministry of Defense, Netherlands.

Zuidweg, J. K. 1970. Automatic Guidance of Ships as a Control Problem. Ph.D. Dissertation. Techn Hogeschool Delft, Netherlands. 136 pages.

D0871833

Thomas Gage in Spanish America

Thomas Gage receiving gifts from his parishioners—a completely fanciful picture
from the first German edition of the *English-American*

THOMAS GAGE
IN SPANISH AMERICA

by

Norman Newton

BARNES & NOBLE, Inc.
NEW YORK
PUBLISHERS & BOOKSELLERS SINCE 1873

First published in 1969
by Faber and Faber Limited
First published in the United States of America, 1969
by Barnes & Noble, Inc.
Printed in Great Britain

S.B.N. 389 01013 8

Contents

5

695713

Illustrations

7

New Spain in the seventeenth century

Preface

THOMAS GAGE was in many ways an unpleasant person. But he was, in most things, a candid one, and it is this relative honesty—not only about himself as a person, but also about the things he saw and heard—which makes him such a good travel writer.

Gage, who came of a distinguished family of English Roman Catholics, was a Dominican friar in 1625, when he volunteered to go to the Philippines as a missionary. In Mexico, however, he decided to run away from the Philippine mission. For twelve years he served as a priest in Guatemala, returning to England in 1637. In England, he left his church, flirted briefly with Anglicanism, and joined the Parliamentarians. His conversion to radical Protestantism, and the influences of a time of moral confusion, seem to have caused what must be called a moral breakdown. His testimony in court was responsible for putting a brother in prison, and sending three priests to their deaths.

In 1648, Gage published a book on his travels, which led to his becoming involved in Cromwell's invasion of Hispaniola and Jamaica in 1655. He died in Jamaica early in 1656.

Gage's book—*The English–American His Travel by Sea and Land, or A New Survey of the West Indies*—appeared at a time when Spanish Mexico, or New Spain, was hardly better known to Englishmen than the Aztec Empire had been. At this time, the long-festering desire of English merchants and statesmen to seize the Spanish colonies in America, or, failing

9

that, to gain control of their economies, was reaching a climax. That 'mechanic fellow', Cromwell (the phrase is Charles the Second's), was in this, though he was so virulent an enemy of many more benign traditions, continuing the foreign policy of Elizabeth the First. Perhaps it would be more correct to say that he was continuing the foreign policy of the Elizabethan merchant class. It was during this time—the Cromwellian interregnum—that the English, who had up to then been forced to content themselves with piracy, laid the foundation of their Caribbean Empire. Gage's book afforded some of the arguments in justification of this venture. In 1655—the year in which Gage himself participated in Cromwell's invasion of Hispaniola and Jamaica—the book was reprinted. And in 1677, Colbert, who had his own 'Western Design', ordered a French translation prepared.

It is not difficult to see why the book should have been so popular in countries which had aggressive designs against the Spanish Empire. It had great propaganda value. It showed that the Spaniards were oppressing the Indians, and that any attempt to 'liberate' them, that is, oppress them oneself, could be justified in the highest moral terms. It was easy enough for the Puritans to forget that their own record in America was, in most respects, much less distinguished. Furthermore, the book was an admirable piece of reconnaissance. Gage seems to have gone out of his way to observe and describe just those things which an invading army would have had to be aware of—roads, harbours, and fortifications. He also noted those sensitive areas of political and religious life which might be expected to harbour possibilities of subversion.

Gage was not a very good writer: his evangelical moralizings sit oddly in a literary style which tends towards a dusty Baroque floridity. But he had marvellous eyes and ears, and a keen sense of social structure, and there is a vivid plain accuracy beneath the somewhat repellent surface of his style. His cool and accurate picture of the country, vivid in detail and precise in its statement of the more general aspects of the culture, reminds one of the water-colours and plans of eighteenth-century army officers.

There seems to be general agreement that, in his descriptions of physical things at least, Gage is to be trusted. Indeed, in most studies of seventeenth-century New Spain, Gage is quoted as an authority, along with Spanish governmental and ecclesiastical documents of the time.

There are certain omissions, of course. Gage makes little mention of the arts of seventeenth-century Mexico, except as examples of the extravagance and wantonness of the society. It should be pointed out, therefore, that this was—considering the size of the population involved—perhaps the most artistically productive period in the history of post-Columbian America. Poets such as Bernardo de Balbuena, Francisco de Terrazas, Fray Miguel de Guevara, Luis de Sandoval y Zapata, and, above all, Sor Juana Ines de la Cruz; playwrights such as Juan Ruiz de Alarcón, arguably the greatest dramatist ever to come out of America; the historians of the Conquest and the pre-Columbian civilizations; the composers Franco, Fructos del Castillo, Bermudez and Padilla; the architects and craftsmen who built the great cathedrals and buildings of state: these created in Mexico a Renaissance and Baroque culture of distinction and nobility. It was, of course, provincial to that of Spain, but its splendour was no less great for that. Furthermore, it was a distinctly American Renaissance and Baroque, in which elements from Europe and pre-Columbian America were joined to form a new whole.

Gage tells a number of stories about events in which he

11

was not personally involved: these are vivid little pictures of Mexican life, and, far from being digressions, they are essential to his picture of New Spain. He was, among other things, a born gossip.

What are we to make of Gage's accounts of the state of religion in America? J. E. S. Thompson, the distinguished United States archaeologist, holds, in his edition of *The English–American* (entitled *Thomas Gage's Travels in the New World*), that they are highly exaggerated and unfair; Gage's other editor, A. P. Newton, seems to accept them as fair accounts of the situation. Certainly, there was much to admire in the Roman Catholicism of Baroque Spanish America, and Gage—anti-Papist as only a convert to Protestantism can be—has nothing to say in favour of Rome. Yet he is careful, by and large, to stick to the facts: all the abuses he recounts can be found in the records of Mexican civil and ecclesiastical courts. To readers nurtured on tales of Spanish cruelties, his account must have seemed rather mild: they would have expected stories of genocide and atrocity, and instead he told tales of venality, greed and deceit—lesser, rather unexciting crimes.

The fact is that, in its stated intentions at least, the Spanish Church showed itself to be as enlightened as any of the other churches of seventeenth-century Christendom. Even the Inquisition, cruel as it was, was not more cruel than the anti-Papist tribunals of Protestant Europe. Nor is the record of the Spanish government dishonourable. A degree of respect was shown towards those native institutions not inextricably bound up with the horrible Aztec religion, and it was held as a point of doctrine that the Indians were, as much as the Spaniards, rational men. But these doctrines were quietly set to one side by the conquistadores, settlers and merchants, who, like other colonial élites in North and

South America, had the rapacity and greed of the new rich. They had, very quickly, won the power and prestige of landed aristocrats, without having inherited that tradition of moral responsibility which had softened the European feudal system. In these qualities, many of the Mexican clergy and friars seem to have differed from the rich land-owners not in kind, but only in degree—in their rejection of the more odious forms of physical coercion, and in the fact that they tended to extort and oppress as corporate bodies rather than as individuals.

Yet the evils of the society—real enough—should not be exaggerated. The Mexican countryside was bountiful in a way which we, used to a rural life denatured and impoverished by the tumour-like growth of the manufacturing industries, can hardly imagine. (The average Indian peasant was certainly better off than the English factory worker of the early 1800's.) The Anglo-Saxon reader, his imagination fed at several removes by reworkings of Las Casas' account of the horrors committed by the Spaniards in the West Indies, tends to think of the Spanish Empire as infamously cruel. In fact, Las Casas exaggerated without scruple: one Spanish scholar, Ramon Menéndez Pidal, has argued convincingly that he was a paranoid, though a well-intentioned one. The first settlement of the West Indies, by a population many of whom were not Christians, and in which criminals, vagabonds and profiteers were more than adequately represented, was certainly accompanied by horrors of every kind; but then so was the settling by Anglo-Saxons of Newfoundland, Australia and California. After this dark period, the Spanish Empire was, in Mexico at least, no more cruel towards the indigenous inhabitants than most empires have been. Indeed, except for certain Creoles 'in advance of their time', the Spaniards were not racially prejudiced in the

modern sense. The mixture of the conquering and indigenous races was the announced aim of Spanish and Papal
policy in the late 1500's. Theirs was a caste, not a racist
society. Gage, who is hardly a friendly observer, finds much
brutal exploitation of Indians and negroes, but he gives only
a few examples of wanton sadism, all of them, as even he
points out, in clear violation of Spanish law.

It is interesting to discover in Gage evidence of the bitter
hostility felt by the Creoles towards the mother country, a
feeling which had become apparent even before the Conquest was completed. A similar feeling was found in the
colonies of Britain and France—one need only mention the
Barbados revolt of 1652, the unrest in Jamaica during the
1720's, the conflict between the merchant seignieurs of New
France and the metropolitan government of France, and
the mercantile dissatisfactions which, fed by dreams of a
New World Order, brought about the revolt of the Thirteen
Colonies. In Mexico, as in the rest of America, a colonial
slave-owning and trading class was restive under the re-
straints—many of them just—imposed upon it by the
government of the home country. Yet curiously, and perhaps this is simply due to the relative prosperity of Mexico,
there were no open separatist revolts, such as there were, for
instance, in Catalonia.

There is one regrettable omission in Gage's book: he never
tells us what he looks like. Thus there moves through his
world of vivid sights, smells and sounds, long since dissipated,
a figure who must be for us, in visual terms, an ambulant
void.

I must say, to conclude this preface, that I have not felt
myself bound by Gage's terminology: for example, he refers,
not to 'Mexico City', but to 'Mexico', a usage which would
only be confusing, since the word now means, to most

English-speakers, the country rather than the city. I have called the capital 'Mexico City' for the sake of clarity, even though this is anachronistic.

Currency comparisons are based on figures given by Gage himself. I have not come across any comparative study of seventeenth-century Mexican and English currency values, or any detailed comparison of relative costs of living. It must be remembered here that a unit of money would have bought much more in his century than it will buy in ours.

I—*The Missionary*

IN less than two months, the galleon fleet would be sailing for America. Friar Antonio Calvo was busy. He had to find six Dominicans in the monasteries of Andalucia, six friars willing to travel halfway round the world, to serve their order, and Christ, in the Philippines. They must be ready to leave Cadiz in the first week of July. In Cadiz, they would join twenty-four friars recruited in the cloisters of Castile and the vicinity of Madrid by the man under whose orders Calvo was acting—the Pope's Commissary for the Dominican Mission to the Philippines, Friar Mateo de la Villa.

Thus, around the end of May, 1625, his quota almost made up from the friaries of Cordova, Seville and San Lucar, Friar Antonio came to the Dominican convent at Jerez. With him he had one Father Antonio Meléndez, of the College of St Gregory in the northern Spanish town of Valladolid.

In that convent, Meléndez was greatly surprised and pleased to discover an acquaintance of his, a one-time fellow student in the College of St Gregory. This was the Englishman Thomas Gage—Friar Tomás de Santa Maria. Gage was then about twenty-three years of age; he was a restless young man, fonder of travel than of study, and less fond of the contemplative life than of either of these. Meléndez took it on himself to persuade the English friar to join the Philippine mission. Since he had been given a good supply of money to cover his travelling expenses, he decided to begin by buying Gage a good supper: the Englishman enjoyed food, and did not much relish friary fare. Not that this was mere policy on Meléndez' part; he was an open-hearted man; he wished to share some of his money with his friend, and his own enthusiasm for the Philippine adventure was quite genuine. The 'good Jerez sack', which, Gage tells us with retroactive disapproval, 'was not spared', inspired him to declaim, almost rhapsodically, about the glories of this country which he had not seen.

He began, decorously and conventionally enough, with the glories of possible martyrdom. One need only be killed by a pagan to achieve eternal glory. Should he, Meléndez, have the good fortune to die converting the Japanese, they would make a book of his life and death; and though he was only the son of a Segovia tailor, the Holy Father would call him St Anthony, and talk of him in the same breath as the Apostles in Heaven.

As the level of the sack crept down the bottle, his rhapsody became somewhat more secular in character, though its tone was occasionally elevated by a reference to Scripture. Had Gage heard, he asked, of the riches of the Philippines? Why, the sails of their ships were not made of coarse canvas, but of Chinese silk! Pearls, rubies and diamonds were there considered mere pretty stones. They paved the ground with tiles of gold and silver. One might consider the islands the true Promised Land: the trees were hung with clusters of nutmegs bigger than the bunches of grapes Joshua and Caleb had brought back from Canaan, and their chocolate, sweetened with sugar from the canes in their fields, tasted like milk and honey.

As the wine-bottle emptied, Meléndez' dream of the Philippines became more and more profane. The land was an Earthly Paradise, a Garden of Pleasures. There every Adam could have his Eve, with no fruit forbidden, for what was a

18

sin in Spain was there only the custom of the country. And wherever a lucky friar might travel, there would be 'Indians' to sound trumpets before him; they would strew flowers beneath the hooves of his horses; they would put up triumphal arches for him; and they would kneel before him as if he were a god.

Will you come with me, Thomas?—he said as they parted, and swore that he would not sleep that night until he knew they would be setting out together. He left Gage with six Spanish pistoles, and a promise that if he could steel himself to face the pleasant martyrdom he had described, Calvo would give him more than enough money to buy supplies for that long and tedious journey, which would take him across two oceans and the breadth of New Spain.

Gage's reply was cautious: he said he would sleep on it. But, he added, he would do much for his old friend. If he did decide to go, he would try to persuade a friend at that convent, the Irish friar, Tomás de León, to accompany him.

Even after one had made allowances for Meléndez' naïve enthusiasm, there was much in the Philippines to attract a young friar anxious to get ahead. Manila was then the centre of trade and missionary activity in Spain's Far Eastern Empire, and in Manila there were two classes of importance, the merchants and the friars. When they were engaged together in the profitable task of spreading civilization and Christianity, it was sometimes hard to tell them apart. Overseas trade was an obsession in that city, which Father Chirino described as 'a copy of that Tyre so praised by Ezekiel'. Agriculture was left to the native Filipinos, and local retail trade and the mechanical arts to the Chinese. Persons of importance in the city devoted themselves to buying goods from the Chinese and Japanese ships which visited Manila each year for the trade fair, and shipping them on, at an appropriate mark-up, to Acapulco. Through Manila came many of those Oriental goods which were becoming essential to the happiness and comfort of the wealthy classes of America and Europe. There were luxury goods, staples and trivia—Chinese silks, fine cottons from India, Persian rugs and carpets, porcelain vessels, spices, gold bullion, jewels, fans, eyeglasses, paper balloons, musk, slaves from South Africa, red lead, tea and copper spittoons.

The religious houses were the banks of the colony, and their interest rates were exorbitant. Practically, this was justified by the dangers of the Pacific crossing; the lingering moral doubts of an age which still retained something of the old fear that usury was a sin against nature were assuaged by the fact that the money the houses earned, in their capacity as commercial banks and marine insurance companies, went to charitable foundations, the *obras pias*, or 'pious works'. The religious orders were almost all-powerful. Rather later than the period in which this book is set, three of the governors of the Philippines were brutally humiliated by the orders: one was first removed from his position, then made to stand every day in the streets of Manila; one was broken by the Holy Inquisition; one was slaughtered by a mob incited by friars.

Gage lay awake most of that night, trying to make up his mind. In the Philippines, he would undoubtedly rise in the Church much more quickly than he would in Spain; he might, indeed, come to live the life of a potentate. On the other hand, he might die at sea, be eaten by cannibals, or die miserably of disease in some wretched cluster of grass huts, stinking of dung and decayed vegetation. In Spain, however, he would probably always remain a poor friar, unless, indeed, he were to be sent back to England as a clandestine priest, to live a hole-and-corner life which might end at any time in the ambiguous triumph of a martrydom. Gage came from a line of martyrs, it is true; but he was the weak member of the family.

He had been away from home ten years or so, having left at the age of about thirteen years to go to the Jesuit college for English boys at St Omer, near Calais. He came from a distinguished and rather dangerous family of country squires, which had risen from obscurity in Tudor times.

He was related, though very remotely, to a number of distinguished persons in the aristocracy and minor gentry, among them Sir Francis Bacon (Lord Verulam), the Earl of Salisbury Sir Robert Cecil, Henry Wriothesley the Earl of Southampton, and also, though the connection was so remote as to be almost meaningless, with William Shakespeare. (In order to avoid misunderstandings, I should hasten to show just how remote this last connection was. The grand-aunt of Robert Southwell, his mother's cousin, was the daughter of Nicholas 1st Lord Vaux, whose other daughter, Catherine, had married Sir George Throckmorton, whose grand-daughter, Mary Throckmorton, had married Edward Arden, son of William Arden, who was third cousin to Shakespeare's mother.)

Gage's great-grandfather had served Henry VIII, when the Church of England had rejected the suzerainty of the Pope, but not the Catholic faith; he had been in disfavour during the Protestant ascendancy under Edward the Sixth, and had risen again during the submission to Rome under Mary. He had been a strong advocate of the marriage of Mary and Philip the Second of Spain, and had taken a leading part in the suppression of the Wyatt Rebellion against Mary. Later heads of the family had followed the same line: they belonged to that body of English Roman Catholic opinion which favoured Spain in feeling.

When in 1570 Pius the Fifth declared Elizabeth the First a usurper, excommunicated her, and released all her subjects from their oaths of allegiance to her, he put the Papists of England in a very difficult position. They could no longer communicate at Anglican altars without being excommunicated, and loyalty to their Queen had become treason to their church. Spain, which was now the chief military and imperialist instrument of the Papacy, and thus the defender of their cause, was also the enemy of their native country.

Thus the Gages became a family of conspirators. They must have believed that they were acting for their country's spiritual good against a régime which, while it had the temporary support of a blinded people, was not only wicked but illegitimate, and thus, in a sense, alien and un-English. Thomas's uncle, Robert Gage, had taken part in the plot of Anthony Babington, in which Spain had taken some interest, to assassinate Queen Elizabeth and replace her with Mary, Queen of Scots. He was captured in a barn and executed in 1586, two years before the Armada was defeated. Thomas's parents, John and Margaret, were almost executed for harbouring Jesuits. They escaped death through the intercession of Lord Howard of Effingham, a distant kinsman, who took, as a reward for this act of charity, their property near Croydon, Haling Park.

Thomas's immediate family were extraordinarily courageous and dedicated in their faith. His father, John, had the fault of his virtue, a rigidity of temperament which would not tolerate weakness. After her own close escape from execution, his mother, Margaret, was brave enough to burst through the crowd lining the street where her cousin, the poet and Jesuit Robert Southwell, was being led to his martyrdom, and to throw herself down before him, asking his blessing. Most of his brothers had distinguished themselves in some way, if only by their devotion to their church. Henry became a soldier in the service of Spain, and was to die fighting for Charles I, who had knighted him; George became a diplomat and a priest; William became a Jesuit. There were two sons by a second marriage: John became a secular priest, and Francis became head of Douai College, in France.

Thus Thomas had passed all his short life in an ambiency of intrigue and danger. His childhood had been spent in a house full of secret passages and hiding-places. He had been present when the solemnity of the Mass, celebrated before a few neighbours and servants, was interrupted by a confused noise of hooves outside the house, a rattle of steel and a loud, brutal pounding on the door. The priest would hastily take off his vestments, and run to a secret hiding place behind a sliding wall panel or the back of a false fireplace. The bread and wine would be whipped away; expressions of piety would change to ones of cheerful nonchalance, and the servants would be sent downstairs to delay the searchers with naïve and bumbling offers of help or honest words of outrage. Nobody could be trusted. One's neighbour, apparently firm in the faith, might suddenly turn around and denounce one; one's oldest and apparently most trustworthy servant might turn out to be a government informer. On the other hand, the agreeable and modest young man, newly arrived in the district, and overflowing with sentiments of loyalty to the régime in power, might reveal himself, at a propitious moment, as a Jesuit, newly arrived from Rome, Spain or France.

One could not even depend upon one's enemies. The anti-Romanist laws, which were very cruel when they were applied with full rigour, were sporadically enforced. By the time Gage was in his teens, there were very few executions, and those receiving lesser punishments seem to have been chosen almost at random, simply to inspire terror in the others.

The Roman Catholics were also in conflict among themselves. There was a virulent subterranean feud between the Jesuits, who were committed to active intrigue, and, if necessary, to assassination and rebellion, and the secular priests, who believed their duty lay simply in ministering to the dwindling body of their co-religionists. The government had nothing to fear from the latter: it often turned a blind eye to private Roman Catholic services, and sometimes intervened against over-zealous local magistrates. But when James the First attempted full toleration in 1603, demanding only that Romanists acknowledge the temporal sovereignty of the King rather than the Pope, hordes of Roman Catholics appeared as if from nowhere. The abolition of recusancy fines resulted in a shrinking of the congregations of many parish churches, and in certain neighbourhoods great numbers began to attend mass openly. Toleration was hastily withdrawn, and, it would seem, with good reason: in 1604 the Gunpowder Plot was discovered. Gage was then about two years old.

It was the policy of the Jesuits to enlist as many promising young Englishmen as they could. When Gage had been sent to school on the Continent—first, when he was only thirteen, at St Omer, and later at Valladolid—he had entered into a strange world which combined a rigid internal and external discipline and the teaching of self-sacrifice even to the point of martyrdom with a most poisonous and subtle form of intrigue. The English government, too, had its counter-intelligence agents in the Jesuit colleges of Europe.

Any personality would have been warped by such a childhood and adolescence; some personalities were terribly warped. It is not surprising that Thomas's thoughts, as he turned over Meléndez' proposition in his mind, were extraordinarily bitter.

He had less reason than most to love his father. He had just received a letter in which his father had informed him that he was disinheriting him because he had joined the Dominican Order. (Gage does not tell us under what circumstances he made the charge.) Not only had Thomas refused to fulfil his father's expressed desire that he become a Jesuit, John Gage said, but he had proved to be, in his affections, 'a deadly foe and enemy unto them'. Furthermore, he had added, Thomas must never expect to see his family again if he returned to England: the Jesuits would see that he was chased out of the country. When he thought of the secret influence exerted by the Jesuits in high places, even at Court, Gage could not underestimate the seriousness of this threat.

It is odd to find Gage, who had taken vows of poverty and belonged to a mendicant order, worrying about his inheritance. But Thomas had never had 'a sense of vocation', any more than those Franciscans of New Spain and Peru, mentioned in a papal declaration of 1578, who went to the New World to make themselves rich, and returned in secular dress. Gage denies it, but there is reason to believe that the factors which were influencing him towards an acceptance of Meléndez' proposition were quite secular in character: his prospects in England had been cut off; he had that deep curiosity about the wonders of America and Asia which then heated the imagination of almost every young man with an adventurous spirit; and he knew that a cunning priest, willing to cut a few moral corners, could do well in the Philippines. After his father died, he reasoned, he could return to England an independent and wealthy man.

That whole night he lay awake, and when the sun rose,

his mind was made up. He met Meléndez at breakfast, and agreed to join the mission. At lunch, he introduced Meléndez to his Irish friend, Tomás de León. And at dinner they met Friar Antonio Calvo, who greeted the pair with, as Gage depicts it, a most unctuous enthusiasm.

Calvo immediately set out to allay one of their fears about the ocean crossing. If they had had to depend on ship's stores, they would have had a very monotonous diet—biscuit, salt fish, salt meat, beans, bacon, cheese, onions and garlic, and what fish was caught during the voyage. But Calvo had a list of the foods he had laid in for his friars—'what varieties of fish and flesh, how many sheep, how many gammons of bacon, how many fat hens, how many hogs, how many barrels of white biscuit, how many jars of wine of Casalla, what store of rice, figs, olives, capers, raisins, lemons, sweet and sour oranges, pomegranates, comfits, preserves, conserves and all sorts of Portugal sweetmeats'. He spoke of his friars in the most flattering terms: he promised to make them all, when they arrived in Manila, Masters of Arts and of Divinity. He took out a heavy purse, and gave them coins to spend that day in Jerez—money to buy whatever they wished, with more than enough left over to carry them to Cadiz, the point of departure. And, to conclude, he spread out his hands, and bestowed a special benediction upon them.

The next morning they set out on the road to Puerto de Santa Maria, on the Bay of Cadiz: from there they would go to Cadiz, where they would meet Meléndez and the other friars. Since their luggage was being sent on by Calvo, they travelled light, riding on little asses 'like Spanish Dons'. As they passed through Jerez, they looked for the last time at its whitewashed houses with their narrow balconies and roof gardens. Then they were out of the town and in the country, where the great vineyards stretched for miles. They jogged on, down the quiet roads, until they came to the banks of the Guadalete, and that most beautiful and noble Carthusian convent which Andrés de Ribera had built some fifty years earlier, its façade ornate with pilasters and urns and the statues of saints.

The Guadalete was thought, in Gage's time, to be the Lethe of mythology, the river where souls drink, and forget the earth. So far as Gage was concerned, it was Spain he wanted to forget, and England as well.

By nightfall, they had reached the estuary of the Guadalete. They were in the town of Puerto de Santa Maria, looking out over the Bay of Cadiz. In the rapidly dimming light they could see, riding in the bay, the great vague forms of the galleons which were to take them to the New World.

They did not have much time, though, to gaze meditatively over the water. They were warriors of Christ, the valiant and humble vanguard of a Spain which, more than any other European country before or since, considered itself specially commissioned to establish a Christian Empire over the entire world. The spirit of the Crusades was still strong in Spain; here the secular and ornate glory of the Renaissance and early Baroque had not smothered the hard heraldic austerity of the Middle Ages, with its mysticism smelling of bread, blood and spring flowers, but had in a sense been swallowed up by it. The missionaries, toreros sallying forth to fight the bull of Hell in heathen lands, were being sainted while they were still alive.

Thus the town fathers of Puerto de Santa Maria had arranged that the streets through which they were to pass would be lined with cheering crowds. Admittedly—and the fact seems to sum up the way in which the realities of seventeenth-century Spanish life tended to betray the ideal—

a large number of the cheerers were galley slaves, commandeered for the occasion. But the cheers were loud for all that, and the trumpets were sounded as if for a procession of kings. When they had run this gauntlet of noisy veneration, one of the town notables, a certain Don Federico de Toledo, entertained them to supper, after which his gentlemen took them to a cloister of Franciscan mendicants, who washed their feet as the Lord had washed the feet of His apostles. They were then shown to bed.

The next morning, after a splendid breakfast at the Franciscan cloister, they were taken down to the shore. There they climbed into boats, and Don Federico's gentlemen took them to Cadiz, where they were to meet their fellow apostles. The Pope's Commissary, Friar Mateo de la Villa, entertained them at his own table, which was 'stored with fish and flesh'. 'There we continued in daily honour and estimation,' says Gage, 'enjoying the sights most pleasant which Cadiz both by sea and land could afford unto us, until the time of the fleet's departing.'

The pleasant sights of Cadiz! Cadiz was then, after Seville, the chief port for the American trade. In one sense it was richer than London or Amsterdam; in another, it was much poorer. It was rich, indeed, in the sense that a great deal of wealth passed through it. Spain, aspiring to be the perfect Christian state in an age of apostasy and venality, was, to its credit, quixotically anti-capitalist. It did not, as did many Protestant countries, attempt to foster the bourgeois way of life as if it were embodied virtue. Indeed, it regulated 'free enterprise' in much the same way that some cities control prostitution. Trade was, certainly, not respectable; those who sought the esteem of their fellows did not seek to make money, but to acquire a coat of arms. The trade of Spain was controlled by merchants resident in other

European states, many of them Jews who had been expelled from the country, and who traded through Spanish front-men. Then, too, Spain was in continual debt: her European wars were expensive, and the bureaucracy was huge. Thus, while Cadiz was at any one time enormously rich, the town was simply a reservoir in which wealth was gathered and stored, so that it could be run out, through various conduits, many of which were secret and hidden, to the merchant towns of Italy and Northern Europe. It was the Genoese, with the exiled Marranos and the Huguenots of France, who profited most from the American trade.

Cadiz was one of the oldest commercial cities in the world: it had been founded by the Tyrians in 1100 B.C., on the site of an even older Tartessian city. Yet, to judge by its appearance, it might have been only a few hundred years old. For thousands of years, men had been pulling buildings down as fast as they put them up: there were no great edifices, and the cathedral, rebuilt after having been destroyed by Sir Francis Drake in 1596, twenty-nine years earlier, was unimpressive. (A few months after Gage left for America, Cadiz would be attacked by another English force, of nine thousand men, led by Lord Wimbledon.)

The town had its charms, though. It was famous for the acuity of its inhabitants, and the seductiveness of its dancing girls. Like Tyre, like Manhattan, it was built upon an island, and the population was crowded into a narrow space. Its colours were white and blue—the shell-white of its houses, so stark and bright in the sunlight that it hurt the eyes, and the pale, burning blue of the sky and sea. Its packed buildings, separated by narrow meandering streets, were as high as six stories. Behind the intricate iron-work of their balconies, red geraniums bloomed in pots. There were view-towers, like tiny minarets, erected on their flat roofs: from these the inhabitants could gaze over the bay to see the first ships of the American fleet, or, perhaps, a flotilla of English pirates. This white, unadorned, baking-hot and crowded city must have produced then, as it does now, odd twinges of disquiet, as if one were out of Europe, and, indeed, out of time. Yet the waterfront would have brought one back to the contemporary realities which Gage's Spain was attempting to deny. There stood the great warehouses of the Genoese men of business—sugar merchants and international bankers closely connected with the financiers of Amsterdam.

It was now approaching the end of June, 1625. The Grand Apostle, Friar Mateo de la Villa, called his thirty Dominican missionaries together. They had thought he was himself prepared for a martyr's death, but he soon disabused them. He regretted that he would be unable to accompany them. His excuses were unconvincing: it seemed clear that, in spite of his words of encouragement, the Grand Apostle would sooner work in Spain for a bishopric than in Japan for a martyr's crown.

When he announced, further, that he had a commission from the Pope to nominate somebody in his place, and that the person in question would be Friar Antonio Calvo, he almost caused a mutiny. Indeed, two of the missionaries ran away. Calvo was an outrageous glutton—a fat bald old man whose white habit was spotted with bacon grease—and he more resembled an ageing scullion-boy than a Papal commissary. Some of the more cynical friars, though, did not object to his appointment. Calvo, they reasoned, was an old simpleton, and could easily be taken advantage of. Furthermore, since he was no lover of austerity, they would at least be sure of eating well.

It was under this slovenly and unattractive person, therefore, that their Apostolic Mission was to proceed—first to

Mexico, three thousand Spanish leagues from Spain, then, three thousand leagues further, to Manila, the viceregal capital of the Philippines.

II—*Sailing to America*

IN the fleet which was to take the Dominicans to the New World there were forty-one ships, including eight galleons for protection against pirates. The ships had already been laden, under the supervision of an official of the House of Trade in Seville. Many of the passengers were on board: some of them, if this fleet was like the others, must have been prohibited immigrants—Protestants, Jews and newly-converted Jews, who were distrusted because they often practised their ancestral rites in private. These prohibited immigrants would have bought their way on board; once they were past the immigration officers of the Spanish colonies, their co-religionists would quietly help them to establish themselves.

There were many smuggled articles. The regulations demanded that all cargoes be registered at the House of Trade in Seville, but foreign merchants, forbidden to trade with the Indies, often put their goods on board by passing them up over the side. In addition, there was a contraband trade carried on by the religious orders: the monasteries and friaries near Seville and Cadiz were virtual smugglers' warehouses. Individual ecclesiastics and officers of the crown, who were not required to pay duty on their personal effects, because they were forbidden by law to engage in trade, often included large quantities of merchandise in their luggage: this would later be sold in America.

The official from the House of Trade had ascertained

31

'whether the ship is capable of making the trip, and the freight which it can take'. He had also checked the ships' manifests, but his inspection had been rather perfunctory by our standards. Each shipper had presented an itemized statement of his share of the cargo, and this had been accepted on face value. To have gone into the matter further would have been to cast doubt on the honour of the shippers, and this would have been outrageous. Nevertheless, the ships' manifests usually bore little relation to the cargo.

The ships themselves were vessels of from three hundred to two thousand tons, and they were too massive, as a rule, to defend themselves well against the small and mobile ships of pirates. Often the merchants put themselves in danger by their own greed: in the Philippine trade, for example, vessels were sometimes poorly armed, because the merchants resented relinquishing any of the precious lading space for the mounting of guns; it was not unheard of for the decks to be piled high with merchandise, while the guns were stored in the hold.

There was little space for the passengers. The cabins were hardly more than closets with lanterns hung before their doors. One had to guard, not only against shipwreck and piracy, but against an acute sense of confinement, and a restlessness which might bring passengers and crews to the edge of violence. Shipboard entertainments were many, but regulations were stern. Gambling, quarrelling and blasphemy were forbidden, as well as drunkenness and the carrying of slave women as concubines. There were places set aside for smoking, as a guard against fire, and here covered pipes and cigars in holders were permitted, but not cigarettes.

It is not surprising that there should have been violations of the rules relating to passengers' personal effects: each passenger was allowed only two small leather-covered trunks, writing materials, a mattress, a pair of bottle-cases for wine, and ten jars of 'chocolates, caramels, sweets, or anything else', which had to be stored under the benches in the cabins which served as beds. These delicacies were almost a necessity, the only relief from the extreme monotony of shipboard diet. There would be fish, of course, and at the beginning of the voyage, there would be fresh meat, fruits and vegetables, but these would not last long, unless one were travelling with a Calvo. On stormy days there would be only cold food and drink; it was too dangerous to light the ovens in bad weather. There were water-jars, some hung in the rigging, and others stored away above or below decks. A few large ships had cisterns. When it rained, water was collected in sloping mats which funnelled the downpour into jars, and sometimes heavy mats were hung in the rigging to collect moisture.

On the first of July in the afternoon, the admiral of the fleet, Don Carlos de Ybarra, gave the order, and the warning gun was fired. The passengers, soldiers and seamen prepared to board ship. Some of the less ardent missionaries were reluctant to leave Cadiz, where they had been a month at liberty. One of them, Friar Juan de Pacheco, who had fallen in love with a Franciscan sister, simply disappeared.

On the next day, the second of July, Thomas Gage received some disturbing news. Pablo de Londres, a 'crab-faced' old English friar in a cloister at San Lucar, near Cadiz, had forwarded to the Governor of Cadiz a letter from 'the Duke of Medina' (probably the very powerful Duke of Medina-Sidonia, whose palace was nineteen miles from Cadiz). It was an order that the Governor search out Thomas, and prevent his sailing. So far as we know, the Duke had no personal knowledge of Gage; he was simply acting on the King of Spain's decision that English Roman Catholics should not be allowed to go to the Indies, since they had a country of their own to convert. Perhaps the initiative had originally come from Friar Pablo, who had apparently troubled Gage before on this matter, sending him several letters urging him to go to England. Friar Pablo had also forwarded, this time to Gage, a letter from the Castilian Provincial of the Dominicans, which made offers of preferment if Gage would forget about the Philippines, and join him in Castile. The Provincial had some knowledge of Gage's family, at least. Under the alias of Friar Diego de la Fuente, he had recently been secretary to the Spanish ambassador to England; thus he knew the situation of English Romanists. But more than that, he had been one of the go-betweens in the secret negotiations aimed at effecting a match between the then Prince of Wales, soon to be Charles the First, and the Infanta Maria of Spain. Thomas's brother, George, had been one of the go-betweens on the English side.

Thomas had no desire, though, to return to England or to go to Castile. Nor did Calvo wish to lose a missionary the day before departure. When the Governor of Cadiz came to Calvo to inquire about Gage, the Englishman had disappeared. He had been rowed to the ship on which the Dominicans were sailing, and hidden in a biscuit-barrel. Later, when all the friars had boarded, the Governor came to the ship; but when he asked if there was an Englishman among them, Calvo resolutely denied it. Jokes about this little deception were common, thereafter, among the company of apostles.

The time of departure had come. The remaining passengers were rowed out in cock-boats to the ship. One by one they stepped nervously on to the rungs of the rope-ladder, and climbed up the side.

Then the sails were hoisted, the anchors hauled in, and
amid shouts, the flapping of canvas, and the running to and
fro of seamen, the ships began to move out. The passengers,
reluctant to go below to their tiny cabins, stood on deck
staring at the shore, and nerving themselves for two months
of danger, deprivation and boredom. 'Adios! Adios!' they
shouted at the watchers on the shore, and the watchers cried
back 'Buen viaje! Buen viaje!' Some of the friars wished they
were ashore again; others hung over the side, aimlessly feeding the fish with bits of cake and pastry given them by the
sisters; others stared about them, marvelling at the forty
great ships, which stood, on the wide lawns of the water, like
palaces and churches whose towers and spires were masts
webbed with rigging and hung with sails. The destinations
of the thirty-three merchant vessels were, to them, mere
names—two for Puerto Rico, three for Santo Domingo, two
for Jamaica, one for Margarita, two for Havana, three for
Cartagena, two for Campeche, two for Honduras and
Trujillo, and sixteen for Vera Cruz. In their holds were
wines, figs, raisins, olives, coarse woollens, linen, iron and
quicksilver for the silver mines of Zacatecas.

Some of the passengers were most distinguished. The year
before there had been a violent quarrel between the Viceroy
of Mexico, the Count de Gelvés and the Mexican Archbishop. The public had become involved; there had been
riots, and the Viceroy had had to take refuge in a monastery.
The King had appointed a new viceroy, the Marqués de
Serralvo; he and his wife, the vicereine, were in the *San
Andrés*. With them was Don Martín de Carrillo, a priest from
the Inquisition at Valladolid, commissioned to inquire into
the causes of the mutiny. In the *Santa Gertrudis* was the new
president of Manila, as well as thirty Jesuits, who had
managed to take passage on the same ship (Gage says) so

that they might ingratiate themselves with him, 'for this cunning generation studies purposely how to ingratiate themselves with kings, princes, great men, rulers and commanders'.

In the *Nuestra Señora de Regla*, there were thirty Mercedarian friars bound for Mexico. Gage and the remaining twenty-seven Dominican missionaries to the Philippines were in the *San Antonio*.

The great fleet cleared the Bay of Cadiz, and, over calm seas, sailed to the Canaries. This was where the galleons, having escorted the merchant vessels through the pirate-infested coastal seas, were to leave them. It was an occasion for high ceremony, and the small boats passed and repassed from ship to ship. The Admiral of the Fleet entertained the Admiral of the Galleons at dinner, a ceremony duplicated, on a lower level of rank, among the captains and officers of the other ships. The passengers hastily finished their letters, which would be taken back in the galleons to be posted in Spain. Then, the dinners and leavetakings over, the Admiral of the Galleons was rowed back to his ship, the guns were fired, and the galleons, their timbers and cordage creaking, pulled away from the merchant ships.

Another tie with Spain had been severed, and the departure of the galleons had much diminished the zeal of some of the missionaries. These tried to persuade Calvo to let them return; but that, of course, was impossible. There was no turning back. Many would never see Spain again.

Yet it was a calm and easy voyage, and the boredom of the long crossing was alleviated by shipboard entertainments. On the last day of July—the day of their patron saint, Ignatius—the Jesuits made the *Santa Gertrudis* as gay as a ship in a tapestry. They trimmed it around with white linen: they hoisted portraits of Ignatius, and flags and banners bearing the arms of the Jesuits. All night there were fire-crackers, and the guns of the ship were fired; trumpets and hymns could be heard across the black water, and the masts and ropes were hung with gently-glowing paper lanterns, which outlined them in beads of light. The next day the Jesuits could be seen moving in procession round the deck of the *Santa Gertrudis*, singing hymns in praise of their patron.

On the fourth of August, St Dominic's day, the Dominicans on the *San Antonio* strove to outdo the Jesuits. Again there were fireworks, gun-fire, and singing by day and night. That day the Dominicans asked all the Jesuits, along with the new President of Manila and the captain of the *Santa Gertrudis*, to dinner on board the *San Antonio*. In the afternoon, after dinner, some of the soldiers, passengers and younger friars acted one of the plays of Lope de Vega, who was then at the peak of his extraordinary career. Gage thought it was as well acted upon the *San Antonio* as it might have been at the Court of Madrid, but when one considers the elaborate nature of Spanish court theatre, it is clear that this statement is an affectionate exaggeration.

On the twentieth of August, 1625, about seven weeks after they had left Cadiz, they sighted their first land. We cannot now identify this island, which the Spaniards then called Deseada. Then there was another island called Marie Galante: Columbus had named it after one of his ships. They passed Dominica, so named because Columbus had discovered it, 132 years earlier, on a Sunday. They were in the Caribbean, in those days still a Spanish sea. Islands were always on the horizon, visible as mere bumps or thickenings of the horizon line. Eventually they came to Guadeloupe.

III—*Guadeloupe*

COLUMBUS had first discovered Guadeloupe in 1493, and in doing so had discovered a predecessor, or a memento of one. In a Carib village, twenty or thirty palm-thatched huts set around a cleared space which acted as the village square, his men had found an iron cooking-pot and the stern-post of a large vessel. They did not recognize the stern-post and the Caribs did not work iron. The name and nation of this particular predecessor of Columbus has remained a mystery. It is possible that these two articles were salvaged from a floating derelict, the wreck of some European vessel which, caught in the trade-winds, had drifted across the Atlantic from Europe or Africa.

When Columbus first visited the island, he found the natives using human skulls as household utensils; there were human limbs curing in the rafters, and human flesh boiling in the pots, along with that of geese and parrots. These natives were the fierce Caribs, who, with the peaceable Arawaks, then inhabited the entire Caribbean—a race of Cains living beside a race of Abels. By Gage's time, however, the Caribs were already rather clownish and pathetic. They had come to depend on the yearly visit of the *flota* as much as the merchants of Vera Cruz. The old Caribbean economy had fallen into ruin: now both Caribs and Arawaks busied themselves, when they felt like working, in preparing sugar-cane, plantains and cured tortoise-meat, which they would give to the Spaniards in exchange for cloth and iron knives. They had become seedy and nondescript, collectors of cultural cast-offs. Their canoes were decorated with the coats-of-arms of the nations who visited them—the Spaniards, English, French and Dutch.

Their canoes slid against the steep-sided ships, and they clambered aboard, figures naked or almost naked, with painted faces, long hanging hair, and thin nose-plates which resembled, Gage notes contemptuously, so many hog-rings. They plucked the Spaniards' sleeves, and fawned on them, begging for wine in sign-language. Soon they were rolling drunk on the deck of the ship.

After the Spaniards had amused themselves, for a while, by watching the Indians, the cock-boat was lowered, so that those who wished could enjoy the pleasures of the island. Some went simply to set foot on dry land, any land; some went to bathe themselves, and to wash their musty and ill-smelling clothes; some went, no doubt, in search of women. Those who disembarked were seamen and lay passengers; a few did not return that night, but slept on shore among the Indians.

The friars disembarked the next morning. Leaving their soiled clothing with some Spaniards whom they had hired as laundrymen, they wandered along the beach alone, or in small groups. Everywhere they were accosted by Indians, who, with the most servile respect, offered them the fruits of the island, and begged trifles in exchange—pins or gloves.

The Indians coaxed the friars to come to their village, so 'we ventured', Gage says, 'to go to some of their houses which stood by a pleasant river, and were by them kindly entertained, eating of their fish, and wild deer's flesh.'

About noon, as Gage and a few companions were walking in the mountains, they came across some of the Jesuits of the *Santa Gertrudis*, engaged in serious conversation with a half-naked mulatto. Gage, when he had gone up to the group, learned from the ensuing conversation that the man was, or had been, a Christian. His name was Luis, and he spoke perfect Spanish. Apparently he had been the personal slave of a gentleman bound for America; but had deserted the ship in Guadeloupe, having found that he preferred the company of Indians to the service of Spaniards. He had hidden in the forest until the ship had gone, and then had emerged, giving the Indians, as presents, certain goods which he had stolen from his master. He had been among the Indians some twelve years: he spoke their language and was married, with three children.

Gage and his Dominican companions joined the Jesuits, who were engaged in a relentless assault on the man's conscience. How could he have forsaken the way of Christ, prostrating himself before rude representations of the spirits of the mountains, waters and forests? How could he have given up the God who had revealed Himself in the full glory of His human nature for a vague sky-spirit—immortal and omnipotent, it is true, but quite uninterested in the affairs of men? Luis, though he had lived as a pagan for twelve years, began to weep. The Gospel is strong. He would gladly go with them, he said, but he could not leave his wife and children, whom he loved. Bring them with you, the Spaniards replied, and their souls will be saved. Furthermore, they added, Luis would be freed, and the Church would support him until he was able to earn a reasonable living. Luis was listening with great attention, until some Indians passed by. When they saw their Luis in earnest conversation with the missionaries, they began to talk angrily among themselves. At this, Luis drew away: it was dangerous, he said, for him to talk to the missionaries; the Indians might kill him to prevent the Spaniards taking him from the island, and, indeed, they would probably turn on the Spaniards.

You have nothing to worry about, said the Spaniards, our

ship has soldiers, small arms and ordnance enough to protect us all. Then they told Luis to bring his wife and children to that place on the shore where the Spaniards were drying their clothes: there would be a boat there to take them on board ship. Luis finally agreed, saying that he would entice his wife and children to the beach with the story that he was going down to trade with the Spaniards. Some of the Jesuits, he said, should be standing by the boat.

The Jesuits went to the shore, to tell the Admiral of their promise to Luis, and to arrange that a cock-boat should be ready as agreed. By now Gage and most of the friars were returning to the ship. When they were on board, a discussion began. Many, motivated either by apostolic zeal or the depressing thought of the many leagues they had yet to travel, wanted to stay on Guadeloupe. Were not the Indians a loving and tractable people? In any case, they said, they would not be in much danger, since the Spanish fleet arrived once a year, and its soldiers would certainly punish the natives if they were mistreated. And after all, added the most zealous among them, the worst that could happen to one would be to find a grave in the belly of a Carib. Had they not come out of Spain to seek the crown of martyrdom?

So they talked, looking at the rich green shore of Guadeloupe, which was still dotted with the small figures of those Spaniards who had not yet returned to the ship. A few black-robed Jesuits were standing near a beached cock-boat, trying not to look as if they were waiting for somebody. Then, inexplicably, everybody began to run towards the water. Many of them were half-naked, but they had left their clothes drying in the sun. The Spaniards were pushing off all the boats, and clambering into them in such haste that they were foundering, or shipping water in dangerous quantities. Some of the women passengers were throwing

41

themselves shrieking into the sea. Arrows were flying out of the jungle like clouds of gnats.

The ship's cannons were fired, and a company of armed soldiers took to the boats and rowed to the shore. When they reached the land, there were no more arrows flying from the trees, and the dead and wounded were scattered about the shore, rocking back and forth in the shallow water near the beach, or floating farther out. Two Jesuits had been killed and ten wounded; three passengers were missing. Fifteen persons had drowned in the attempt to escape. Among the wounded was one Friar Juan de la Cueva, who, some short time before, had been trying to persuade Gage to stay on shore a little while longer. He had an arrow in his shoulder.

Luis had not turned up. It was assumed that he had told the Caribs of the Jesuits' plan to steal him away, or that the Caribs, seeing him in conversation with them, had made him confess it. But the first explanation seemed more probable to Gage: Luis had told the Jesuits he would know them by their black coats, and it was observed that most of the arrows were directed at the Jesuits, five of whom were struck in slightly more than fifteen minutes.

All that night the soldiers kept watch along the shore. An armed guard accompanied those who went to the river to fetch water; and muskets were fired at irregular intervals, both to frighten the Indians, and to attract the attention of the three passengers who were missing in the jungle. But the Indians had vanished, and so, it seemed, had the three passengers. A watch was mounted on board the ships and few slept, for fear the Indians might attack in their canoes. Friar Juan de la Cueva lay sleepless and in agony, the wound in his shoulder festering.

At noon, with a good breeze, they hoisted the sails, and left Guadeloupe.

Most of the friars now began to lose their desire for martyrdom, fed by books of saints' lives and monastery frescoes. It was the first time, doubtless, that many of them had seen death by violence. Friar Juan de la Cueva was in continual pain, and his body was swelling: the arrow, it was clear, had been poisoned. On the twenty-second of August, four days out from Guadeloupe, he died.

'His burial', says Gage, 'was as solemnly performed as could be at sea. His grave being the whole ocean, he had weighty stones hung at his feet, two more to his shoulders, and one to his breast; and then the superstitious Romish dirge and requiem being sung for his soul, his corpse being held out to the sea on the ship side with ropes ready to let him fall, all the ship crying out three times, *Buen viaje, a good voyage*, to his soul chiefly, and also to his corpse ready to travel to the deep to feed the whales: at the first cry all the ordnance were shot off, the ropes on a sudden loosed, and Juan de la Cueva with the weight of heavy stones plunged deep into the sea, whom no mortal eyes ever more beheld. The like we saw performed in the ship of *Santa Gertrudis* to another Jesuit, one of the three who had been dangerously wounded by the Indians of Guadeloupe; which likewise died like our friar, his body being swelled as with poison.'

The friars were now discouraged and frightened, and Calvo tried to comfort them. The natives of the Philippines, he said, were not like those of Guadeloupe. Most of them were Christians already, and treated the priests as if they were gods; and those who were pagans stood in awe of the soldiers.

They passed Puerto Rico, and then the great island of Santo Domingo or Hispaniola, long settled by the Spaniards. Now many of the ships began to leave the fleet, heading for ports in the Caribbean or Central America. Only the sixteen ships of the Mexican fleet were left.

In the Gulf of Mexico they were becalmed, and for a week

43

the ships lay, barely moving, in the blue and tepid water. The friars passed the time angling for the little golden fish known to the Spaniards as *dorados*. But the heat was unbearable: the unpitying sun glared on the decks and pitchy sides of the ships, and their sweat ran continually. The evenings and nights were only slightly more comfortable. Nobody could sleep in the cabins. They were, after all, as tiny as the fo'c'sle of a modern tug boat: the ceilings were too low for standing; the beds were only little benches attached to the sides of the ship, with a few cushions for comfort; they smelled of tar and unwashed bodies, and were hung with musty clothes, water-jars and miscellaneous gear. At night the friars, stripped to their shirts, walked, sat or lay on the deck. The mariners washed frequently to cool off; some of them swam by the sides of the ships.

'The nearer we come to the mainland,' says Gage, 'the sea abounds with a monstrous fish, called by the Spaniards *tiburón*.' He had seen one caught with an iron trident: 'a most monstrous creature, twelve ells long at least'. Since these sharks stayed away from the ships, one could swim by the sides without much danger. But one of the mariners of the *San Francisco* was something of an exhibitionist: he began to swim from his ship to another, where he had friends. A shark pulled him three times under the water, taking off a leg, an arm and part of a shoulder. By the time they had lowered a boat, he was already dead. They picked up the bleeding trunk, and buried it as they had Friar Juan de la Cueva.

Three days later, the wind rose, and they moved out of the calm.

IV—*Arrival in Mexico*

IT was seven o'clock, on a Monday morning, and one of the friars was saying Mass. Then a voice cried out *Tierra, tierra*. The communicants rose from their knees to see the American mainland.

At about ten that morning, they could see the coast, and they were running towards it under full sail. Friar Calvo was slaughtering some of his best fowls. Their anguished and strangled clucking mingled with the chatter of the passengers.

The harbour of San Juan de Ullua, which was also called Vera Cruz after a now-abandoned port to the south, was difficult to enter. There were a number of reefs and banks of shingles: one of these, the Arrecife de la Gallega, which lay about a quarter of a mile from the coast, was high enough to shelter the harbour from the dangerous northerlies. On this reef a great fort had been built, to further shelter the harbour from pirates, and beyond it, in the harbour, flags and buoys marked scattered rocks. The town was built on the shore.

It was September, and the season of the northerlies: there was thus some danger of being driven on to the rocks. The pilots, the navigating officers of the ships, met in council, and they decided that, since they would have the help of the harbour pilots the next morning, they would lower all but their middle sails, and approach the shore slowly during the rest of that day and night. That night, a double watch was mounted.

Just before midnight, the wind turned to the north. This change, even though the wind was not at all boisterous,

45

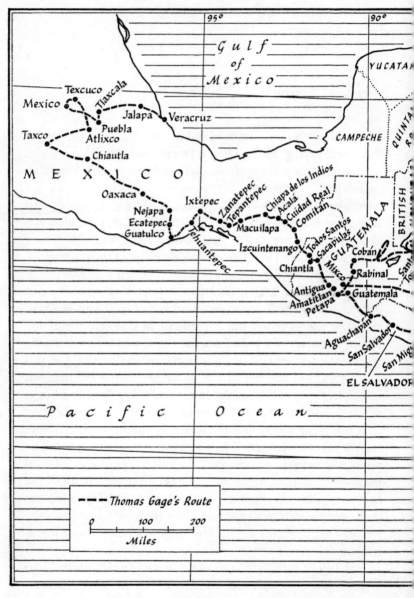

Travels of Thomas Gage in Mexico and Guatemala (1625–37

85° 80°

20°

C a r i b b e a n   *S e a*

15°

Trujillo

O N D U R A S

Comayagua

NICARAGUA

El Viejo
Realejo

León
Granada

Lake
Nicaragua

Salinas
Bay

Rio Desaguadero
Rio Sueré

Nicoya

10°

Cartago Portobello

COSTA RICA Panama

P A N A M A

through Central America to Portobello (1637)

created a mild panic. The friars lit candles, and all night they sang hymns and litanies, and offered up prayers to the Virgin. 'Towards the dawning of the day', says Gage, 'our wonted gale' [presumably it was an easterly wind] 'began to blow again.'

At about eight in the morning, they came in sight of the Gallega Reef, with its fort, church and hospital. The houses of Vera Cruz lay beyond. They signalled for boats; and as the boats were launched from the shore and moved towards them, the musicians on the ships began to play their oboes and trumpets, the guns were fired, and the guns on shore replied.

Cautiously the ships manœuvred round the fort and into the shadow of its great wall, which rose, green with seaweed, from the sea; the stern anchors were run out, and the head-ropes secured to the huge bronze rings in the wall.

If Gage had been standing on the poopdeck, looking over the stern of the ship, he would have seen, among the craft putting out from the shore, the boats of the Crown customs officers and inspectors of the Holy Office. Mexican customs officials and immigration officers may have been, by modern standards, relatively unsuccessful in their attempts to keep out smuggled goods and subversive persons; but they did try.

The town of Vera Cruz, so famous for its riches, was in fact ugly and down-at-heel. The town-site, surrounded by the coarse lush vegetation of the coastal jungle, was a stretch of dirty sand broken by tiny oozing streams and pitted with bogs, 'which, with the great heats that are there, cause it to be a very unhealthy place'. Vera Cruz did not have the wharves of a modern port, since ships anchored by the fort. A simple mole stuck out into the water, with a Customs House at its landward end. The town itself was a straggling assemblage of crudely-constructed buildings, which were, though larger,

hardly more substantial than booths at a fair. Indeed, that is
what they were: like Acapulco, and Portobello (which Gage
would visit later), Vera Cruz was the setting of a great annual
market. The Spanish *flota* had much in common with the
Asian or medieval European caravan, and, like the caravan,
disposed of its goods at a fair. The rich merchants met the
fleet at Vera Cruz; there they collected the goods consigned
to them or haggled over goods sent for resale in Mexico or the
Far East; and when the fair was over, they left. Their homes
were in such inland towns as Jalapa: men of position, able
to make a choice, would hardly live in such a hot, dank,
disease-ridden town as Vera Cruz. The permanent residents
were poor and ignorant—peasants, fishermen, beach-com-
bers. Gage could have seen them on the beach, gazing out at
the ships: Negroes, Indians and poor whites, almost naked
or dressed in rags. Here and there were lumbering ox-carts,
and donkeys half-hidden beneath loads of faggots.

But now there were a number of splendid personages on
shore: wealthy merchants, pompous government officials,
and priests. The Inquisitors and customs officials, already
aboard the ship and checking the passenger lists and ships'
manifests, had their own ominous authority.

When the friars and Jesuits reached the shore, some of
them kissed the ground, sanctified by the labours of the first
missionaries, and others knelt to offer up prayers to the
Virgin and the saints. They then went to stand with their
fellows in the places appointed to their respective orders.

'In the meantime all the cannon playing both from
ships and castle, landed the Viceroy and his lady and all
his train accompanied with Don Martin de Carrillo the
Visitor-General for the strife between the Count of Gelves,
the last Viceroy, and the Archbishop of Mexico. The great
Don and his lady being placed under a canopy of state,

began the *Te Deum* to be sung with much variety of musical instruments, all marching in procession to the Cathedral, where with many lights of burning lamps, torches, and wax candles, was to the view of all set upon the high altar their God of bread; to whom all knees were bowed, a prayer of thanksgiving sung, holy water by a priest sprinkled upon all the people, and lastly a Mass with three priests solemnly celebrated. This being ended, the Viceroy was attended on by the Chief High Justice, named *Alcalde Mayor*, by the officers of the town, some judges sent from Mexico to that purpose, and all the soldiers of the ships and town unto his lodgings. The friars likewise in procession with their cross before them were conducted to their several cloisters.'

V—*Vera Cruz*

GAGE and the other Dominicans—Friar Calvo at their head —were marched to the Priory of St Dominic. The Prior welcomed each of them with a cup of chocolate and some sweetmeats; this was followed by a sumptuous dinner of capons, turkey cocks and chickens—an example, he said, of the abundance of the country. After dinner he entertained them with songs, accompanying himself on the guitar.

Gage professes himself to have been more shocked than pleased. The Prior particularly enraged him. He was not a grave and grey-headed man, such as one would expect to find governing young and wanton friars; rather, he was a gallant young spark, feather-headed and shallow of temperament, who had purchased his position from his Provincial for a thousand ducats. His private chamber contained only a few dozen decayed volumes of theology, and they were covered with dust and cobwebs; but his guitar was hung on the wall within easy reach. Round about the chamber were hung many pictures and tapestries in the style of the country, made of cotton or the feather-work of Michoacán; his tables had covers of silk and his cupboards were full of dishes and cups from China, all of which, Gage tells us (he had a sweet tooth, and noticed such things), were filled with comfits. His conversation was neither pious nor edifying. He boasted a good deal—about his birth, his cleverness, the fondness his Provincial had for him, the love borne him by the wives of the rich merchants, and his great skill in music. When he sang, his voice quivered artfully and lasciviously, and the song was one he had composed in praise of 'some lovely Amaryllis'. Gage, munching savagely on sweetmeats flavoured with musk and civet, noted it all for future reference.

Eventually the coxcombical Prior ran out of songs and self-praise. Gage and the other friars decided, since they would be there for only a day and a half, to take a turn about the town. It took them only an afternoon to see all there was to see.

'Of the buildings,' Gage says, 'little we observed, for they [are] all, both houses, churches, and cloisters, built with boards and timber, the walls of the richest man's house being made but of boards, which with the impetuous winds from the north hath been cause that many times the town hath been for the most part of it burnt down to the ground.'

The population of Vera Cruz was, in Gage's time, about three thousand, and there were among them (though probably not permanently resident in the port) 'some very rich merchants, some worth two hundred, some three hundred, and some four hundred thousand ducats'. A few of the houses, like those of Cadiz, had watch-towers, so that the nervous money-bags who owned them could look out to sea, and be warned of the approach of English or Dutch pirates. Others were connected by underground passages. The townspeople were never free from the fear of attack by privateers or pirates, since it was here that much of the silver and gold of the Indies, and many of the rare goods of China and Southeast Asia awaited trans-shipment to Spain. But the town itself was open: Vera Cruz trusted for its defence to the difficult harbour entrance, and the fort on the Arrecife de la Gallega.

That night, when the friars were sleeping, a strong north wind blew up. The rough boards of the Priory began to rattle together, and the entire building was tottering and shaking so much that the newly-arrived Dominicans, fearing its imminent collapse, ran barefoot into the dirty yard. The resident friars laughed at them: they never slept easier, they said, than when the wind was blowing; it kept raiders away, at least. But in the morning they found that the wind had snapped the cables of one of the ships in which they had crossed the Atlantic, and driven it out to sea.

They had no desire to stay in the unhealthy little port, and they were not disappointed when Calvo announced that they were moving inland the next morning. Thirty mules, brought from Mexico City six days earlier, were already waiting for them. Calvo, who seems to have been a conscientious and thrifty quartermaster, oversaw the repacking of the provisions that were left over from the voyage—wine, biscuits, bacon, salt beef, twelve live hens and three sheep. That evening, they went to the church to see a comedy which the town had prepared for the entertainment of the Viceroy, and the next morning they set out for Mexico City.

VI—*From Vera Cruz to Mexico City*

IN the journey from Vera Cruz to Mexico City, Gage was for the first time treated in the manner Meléndez had led him to expect. The priests and friars were entertained as if they were visiting gods, or Oriental despots at the very least, that is, they were greeted with that elaborate ceremonial humility with which the Mexican peoples had once approached the lords and high priests of the Aztecs. The first town they came to was Old Vera Cruz, now a small Indian village. Twenty of the important villagers met them on horseback: they gave each of the friars a nosegay of flowers, and rode a bow-shot ahead of them until they were met by the chief party from the village—trumpeters, choristers, and heads of the various saints' sodalities.

The procession moved to the chief market-place of the town, and after the municipal officers had knelt and kissed the hands of the friars one by one, they rose to deliver elaborate and interminable speeches of welcome, calling them the Apostles of Jesus Christ, saying they honoured them as if they were gods on earth, and thanking them for having crossed the sea to save their souls. While they were listening, the friars drank chocolate. The spokesman for the friars responded by saying that they indeed cared for nothing but saving the souls of the Indians, and that it was for this they had faced unimaginable dangers by land and sea.

This exchange of florid civilities went on for about an hour. Then the earthly gods took their departure, after having given the village notables some beads, medals, brass crosses

53

and religious relics from Spain, and granted everyone in the village an indulgence of forty years. The market-place was full of kneeling Indians, and the friars, as they departed on their mules, lifted their hands, scattering blessings on their bowed heads.

The trumpeters and choristers, lustily blowing and singing, accompanied them a mile out of town, then left them in order to return to their village. The friars went on through the fertile and varied countryside, which—with its great canyons and tree-carpeted barrancas, its rich pastures, corn-fields and sugar-plantations climbing up the sides of steep hills, its valleys where the vegetation proliferated with jungle-like fecundity, its mountain peaks marching upwards to the interior—resembled a Romantic painter's dream of the Garden of Eden. Yet everywhere Gage found evidence of luxury, extortion and venality.

In the town of Jalapa they stayed in a huge Franciscan convent, where six friars were living on the revenues, ade-quate to support twenty men in excellent style, of a huge monastic estate. The rules of their order demanded that they wear sackcloth and shirts of coarse wool, and that they go bare-legged, shod with wood or hemp; but these friars wore beneath their habits (which they sometimes tucked up to the waist, the better to display such splendour), shoes of fine Cordovan leather, fine silk stockings, drawers with three inches of lace at the knee, holland shirts, and doublets quilted with silk. They were fond of gambling, and acquainted with gamblers' oaths. Each, no doubt, had his own way of re-conciling such luxury and greed with the oaths he had taken. One, whom Gage observed in a game of primera, jokingly used the end of one sleeve to sweep his winnings into the other, explaining that the rules of his order did not allow him to touch money.

There is no doubt that the worldly opulence of the Mexican church was well in excess of the splendours necessary to ritual and the income necessary to use: we must not take Gage's criticism as a simple expression of prejudice. As early as 1556, the Archbishop of Mexico, himself a Dominican, had complained of convents, intended to accommodate only two or three friars, which 'would more than suffice for Valladolid'. The labour, and often the monetary costs, of construction and upkeep were supplied by the Indians.

They passed through La Rinconada, an inn in a low hot valley where they were tormented by gnats, and through Segura, now called Tepeaca. It was in Segura that Gage first tasted the zapote, whose sweet brown or scarlet-purple flesh tastes more like a man-made confection than a fruit. They came to Tlaxcala, whose inhabitants, joining with Cortés against their traditional enemies, the Aztecs, had made the Conquest possible. In his book, Gage, who seems to have made up for his gluttony by his vigorous disapproval of the pleasures of art, looks back with distaste to the music he heard in the church attached to the Franciscan cloister of Ocotelulco, one of the quarters of Tlaxcala. Here there were some fifty Indian singers, organists and instrumentalists, who, Gage says, 'set out the Mass with a very sweet and harmonious music, and delight the fancy and senses, while the spirit is sad and dull as little acquainted with God, who will be worshipped in spirit and in truth'.

From Tlaxcala they went to Puebla, then, as now, one of the most beautiful cities of America. Built on the site of the insignificant Aztec village of Cuetlaxcoapán, it was essentially a colonial foundation, with no Indian past of importance. Indeed, the legend of its foundation (Puebla is full of legends of the colonial period) states that the site was shown to its founder in a dream. Fray Julián Garcés, the story goes, saw

55

two angels holding a string and a measuring rod over some land at the foot of Popocatépetl: this is why the city subsequently built on the site was called Puebla de los Angeles, the City of the Angels. Gage's account, more prosaic, merely says that the city was built in 1530, by order of the Viceroy. It has been called, from its more than sixty churches, 'the Rome of America': in Gage's time, too, it was notable for its many ecclesiastical buildings. The noble cathedral, based on designs by Juan de Herrera, the architect of the Escurial, was then in construction; it would be finished in 1649, when Gage was back in England.

'This city is now a bishop's see, whose yearly revenues since the cutting off from it Jalapa de la Vera Cruz are yet worth above twenty thousand ducats. By reason of the good and wholesome air it daily increaseth with inhabitants, who resort from many other places to live there; but especially the year 1634, when Mexico was like to be drowned with the inundation of the lake, thousands left it, and came with all their goods and families to this City of the Angels, which now is thought to consist of ten thousand inhabitants. That which maketh it most famous is the cloth which is made in it, and is sent far and near, and judged now to be as good as the cloth of Segovia, which is the best that is made in Spain, but now is not so much esteemed of nor sent so much from Spain to America by reason of the abundance of fine cloth which is made in this City of Angels. The felts likewise that are made are the best of all that country; there is also a glass house, which is there a rarity, none other being as yet known in those parts. But the mint house that is in it, where is coined half the silver that cometh from Zacatecas, makes it the second to Mexico; and it is thought that in time it will be as great and populous as Mexico. Without it there are many gardens, which store the markets with provision of salads, the soil abounds with wheat, and with sugar farms, among the which not far from this city there is one so great and populous, (belonging to the Dominican friars of Mexico) that for the work only belonging unto it, it maintained in my time above two hundred blackamoor slaves, men and women besides their little children.'

What Gage does not tell us is that Puebla had achieved its reputation as a cloth town at the price of much human suffering. The mills, in which Indians, Chinese, Filipinos and Africans worked along with convicts serving their sentences, were quite as bad as those of early nineteenth-century England, and were thus, in a sense, 'progressive'. Hours were cruelly long, and the workers, many of whom were children of six or seven years old, were often locked in the factories overnight. It is true that the labour laws of the Spanish Empire, which were comparatively just, forbade such practices; but social justice was already coming to be considered a reactionary concept: the trend was towards unrestricted free enterprise, and it was becoming more and more difficult to enforce laws which the mercantile and manufacturing classes considered punitive.

Puebla was thus an important manufacturing and business town, whose substantial citizens would no doubt have approved the following two stanzas from *La Grandeza Mexicana*, by the former Abbot of Jamaica, and at that time Bishop of Puerto Rico, Bernardo de Balbuena—one of the first to attempt to make poetry out of the odious paradoxes of enlightened self-interest—

> *There is no place where greed may not be found.*
> *Each occupation has its dividend.*
> *Cupidity makes the whole world go round,*
>
> *And all things does he strengthen and defend.*
> *He is the sun who vivifies the ground,*
> *And governs the beginning and the end.*

57

(That very year, though, the progressive Bishop was to suffer at the hands of the chief exponents of free enterprise in that age: in 1625, the Dutch occupied San Juan de Puerto Rico, and burned down his house.)

From Puebla they went to Huejotzingo. Here Gage was profoundly shocked to discover that the Franciscans were teaching the Indian children to dance to the guitar. The friars were entertained till midnight by these children, who sang Spanish and Indian tunes, 'capering and dancing with their castanets, or knockers, on their fingers'.

Then, from Huejotzingo, they climbed the mountains around Mexico, and looked down on that famous valley, which had been civilized almost as long as Spain. Here there had stood the great imperial city of Teotihuacan, a city of eight square miles, larger than Imperial Rome. Then had come the empires of the Toltecs, the Chichimecs, and the Aztecs—all centred in this valley. Finally, new Spanish-speaking cities and a Baroque Christian culture had obliterated the cruel society of the Aztecs, leaving only a sullen folk-culture like that of the peasants of the outlying parts of Europe.

On the third day of October, 1625, they entered the city. But they passed through it, since they were on their way to the Dominican establishment among the gardens near Chapultepec, known as San Jacinto. Gage was to stay there five months.

VII—*The City of Mexico*

THE house at San Jacinto was a beautiful place. Its grounds, in which there were grape arbours and shady walks under the orange and lemon trees, covered fifteen acres. All the fruits of Mexico were to be found there—pomegranate, fig, plantain, zapote and pineapple. Discipline was minimal, and life was undemanding. There was good reason for this apparent laxity. The friars had to stay in Mexico for a few months, if for no other reason than that the fleet did not sail from Acapulco till February. Travel was hard in those days, and they needed a rest between the long journey from Spain to Mexico City, and the much longer journey from Mexico City to the Philippines. Furthermore, since the life of even an average Mexican religious was much freer and more luxurious than that any of the Spanish friars had been used to, many of them would have been tempted, had they been condemned to five months of austerity, to desert the Philippine mission and take refuge in a Mexican religious community. Thus, in the house at San Jacinto, which was under the control of the Dominican superior in the Philippines, they were consciously pampered.

Living under this régime of luxury, Gage's appetite seems to have increased. Every Monday morning, each of the friars was brought half a dozen boxes of conserves of quinces, and other sugared fruits, as well as a good supply of biscuits. Gage found that he was continually nibbling between meals, and he asked a physician to explain this oddity. The physician, who seems to have been of a type Molière (had he not been only three years old at the time) would have recognized, replied that while Mexican meat looked as good as Spanish, it was far inferior because the pasture was drier and scantier. Secondly, he said, 'the climate of those parts had this effect, to produce a fair show but little matter of substance', thus both flesh and fruits 'have little inward virtue or nourishment at all in them'. Gage draws a moral from this: 'I have heard reported much among the Spaniards to have been the

answer of our Queen Elizabeth of England to some that
presented unto her of the fruits of America, that surely where
those fruits grew, the women were light, and all the people
hollow and false-hearted.'

Since San Jacinto was only two miles from the City of
Mexico, Gage often went to town for the day, no doubt
following the road to the point where the stone aqueduct,
which carried water to the capital from the springs at
Chapultepec, turned into the outskirts of the city.

Beside the high arches of this aqueduct were neat and
fertile squares of farmland, through which meandered a
sluggish canal, thick with weed. There were seven of these
canals crossing the city, their noisome waters flowing beneath
pretty little bridges. Had they been kept clean, and had the
water circulated more freely, no doubt these canals would
have given the city a charm similar to that of Amsterdam.
They were too narrow, though, for the density of the freight
and passenger traffic which used them. Furthermore, they
were, at certain times of the year, almost dry, and since there
was no adequate system of garbage collection (although,
Gage says, 'scavengers' were appointed), they received all the
filth of the town. The Spaniards had not followed the
example of the Aztecs, who, thrifty as Chinese peasants, had
carried away the manure, night-soil and garbage of the city
to be used as fertilizer.

Mexico City was quite large: there were between thirty
and forty thousand Spaniards, eighty thousand Indians and
mestizos (Gage much underestimates the size of this element
of the population), and fifty thousand negroes and mulattoes,
both slaves and freedmen. Thus, its white population alone
was as great as that of the Bristol or Norwich Gage knew, and
its total population of one hundred and sixty-five thousand
(the Indians, mestizos, Negroes and mulattoes making up

the lower, serf and slave classes) was nearly two-thirds that of London in 1603. It had a circumference of six miles. Yet, compared to the island-city, Tenochtitlan, it was a small town: exact population figures for Tenochtitlan are not available; but one reliable estimate gives the city itself a population of between five hundred and sixty thousand and seven hundred thousand, and what might be called Metropolitan Tenochtitlan, including the satellite villages, a population of over one million.

Nevertheless, it was, in many respects, a magnificent city. Beneath the hard gem-like blue of the plateau sky, there rose the towers, steeples and tiled domes of a colonial capital which had arisen in an age obsessed by the visual images of grandeur. There were fifty or more churches, convents, monasteries and hospitals; there were the ornate government buildings of stone and brick, and the low, wide-spreading houses of the rich, with their patios and flowers. There were the tidy shops of the city centre. But there were also the grim *obrajes*—workshops and factories—and, scattered in the suburbs, the wretched huts of those who had originally owned the land, and now supported it on their naked backs.

As to the houses of the wealthy, some were two storeys high, with French roofs, but most were in the Mediterranean style—low, extended and flat-roofed. Their walls rose beside the street; the eaves of wood or pottery would, in the rains of summer, pour torrents of water on to the road. They were built around courtyards and patios, invisible from the road, which contained pools, statues and flowering trees; and behind each house was a corral for horses and other domestic animals.

Their gates and doors opened directly on to the street, and many of them were deep-carved with coats of arms. Most of

61

the Creole aristocracy were *hidalgos,* or 'gentlemen', the lowest rank of the Spanish upper class. Some, it is true, were counts or marquises. As a rule, though, most Creole titles had been purchased; thus, while they may have been impressive enough in the New World, they were scorned in the Old.

'Their buildings are with stone and brick very strong, but not high, by reason of the many earthquakes, which would endanger their houses if they were above three storeys high. The streets are very broad, in the narrowest of them three coaches may go, and in the broader six may go in the breadth of them, which makes the city seem a great deal bigger than it is. In my time it was thought to be of between thirty and forty thousand inhabitant Spaniards, who are so proud and rich that half the city was judged to keep coaches, for it was a most credible report that in Mexico in my time there were above fifteen thousand coaches. It is a by-word that at Mexico four things are fair; that is to say, the women, the apparel, the horses, and the streets. But to this I may add the beauty of some of the coaches of the gentry, which do exceed in cost the best of the Court of Madrid and other parts of Christendom, for they spare no silver, nor gold, nor precious stones, nor cloth of gold, nor the best silks from China to enrich them. And to the gallantry of their horses the pride of some doth add the cost of bridles and shoes of silver.'

And clothing, Gage tells us, was equally extravagant—

'Both men and women are excessive in their apparel, using more silks than stuffs and cloth. Precious stones and pearls further much this their vain ostentation; a hat-band and rose made of diamonds in a gentleman's hat is common, and a hat-band of pearls is ordinary in a tradesman; nay, a blackamoor or tawny young maid and slave will make hard shift but she will be in fashion with her neck-chain and bracelets of pearls, and her ear-bobs of some considerable jewels. The attire of this baser sort of people of blackamoors and mulattoes (which are of a mixed nature, of Spaniards and blackamoors) is so light, and their carriage so enticing, that many Spaniards even of the better sort (who are too, too prone to venery) disdain their wives for them.'

Just before it reached the fashionable part of town, the aqueduct turned towards the convent of San Francisco. This was one of the great ecclesiastical complexes of the city: with its houses, gardens, hospital, monastery and church, it covered the equivalent of four modern city blocks. The viceroy attended church here, and here, in 1629, would be reinterred the bones of Hernan Cortés, the conqueror of Mexico. The church itself was built of stones from the great central pyramid of Tenochtitlan.

To the west of the Convent of San Francisco was a poor quarter, containing the College of San Juan de Letrán for mestizo boys, and another college for mestizo girls. Nearby, on marshy ground, were some wretched Indian hovels: they could not be seen from the main thoroughfare, since—to quote one Mexican writer's scornful exaggeration—they were built so low they scarcely seemed to lift themselves from the ground. They were scattered without order, the writer says, 'as is the ancient custom among them'. This area seems to have had something of the quality of a depressed Indian village, set on the outskirts of a large city. In this miserable quarter were a few shops, some of them owned by wealthy persons who, of course, lived elsewhere in the city, a public market, and a notably high gallows, built in a tower, and reached by a door and staircase.

Most of the Indians of the urban area had been forced into quarters like this one. Some of the descendants of the Aztec ruling class had become a semi-captive indigenous gentry, and were entitled to be called 'Don'. These lived with some dignity, their position being higher than that of their fellow mestizos and Indians, though below that of the Spanish gentry. The mestizos, as a rule, were in a lower middle class position: a mestizo would be considered successful if he had become a shop-keeper or a steward. The urban Indians, though, were being driven deeper and deeper into indigence. As Gage expresses it, 'The Spaniards daily cozen them of the small plot of ground where their houses stand, and of three or four houses of Indians build up one good and fair house after the Spanish fashion with gardens and orchards. And so is almost all Mexico new built with very fair and spacious houses with gardens of recreation.'

Indian education, conducted by the earliest missionaries on a high level of idealism, had deteriorated; and the time was fast approaching when persons of mixed blood would be barred from the universities, because it was claimed, there were too many of them appearing in the professions. It was difficult enough, if one was an Indian, to be a 'poor but honest' workman. In most of the guilds, which were poisonous with lower-class white prejudice, an Indian could not hope to advance beyond the status of a journeyman. Indeed, it was hard for him to be accepted as an apprentice. Usually he was hired as an unskilled labourer. The craftsmanship of the Indian was universally recognized by persons of taste, yet he had to work for very low pay, and without the protection of the guilds, who feared his competition.

San Francisco was only one of the churches whose great domes rose above the confusion of lesser domes and spires. In the city's marshy southern quarter, encircled by private homes, there was San Agustin. It had just been rebuilt for the third time, its two predecessors having, says Gage, 'quite sunk away'. In the northern quarter, in a large enclosed square, there stood the church of Santiago Tlaltelolco. Not too far away was Santo Domingo, which had been surrounded, seventy-five years earlier, by a number of tall old houses:

Manila in 1660, from a drawing by Johannes Vingboons

Rural scene in the Philippines

Guadeloupe, as seen by Samuel de Champlain in 1599

here had lived the pioneer élite of Mexico City, the first descendants of the conquistadores.

The social centre of the city was the Alameda; and Gage's description of it is particularly vivid—

'The gallants of this city shew themselves daily, some on horseback, and most in coaches, about four of the clock in the afternoon in a pleasant shady field called *la Alameda*, full of trees and walks, somewhat like unto our Moorfields, where do meet as constantly as the merchants upon our exchange about two thousand coaches, full of gallants, ladies and citizens, to see and to be seen, to court and to be courted, the gentlemen have their train of blackamoor slaves, some a dozen, some half a dozen, waiting on them, in brave and gallant liveries, heavy with gold and silver lace, with silk stockings on their black legs, and roses on their feet, and swords by their side; the ladies also carry their train by their coach's side of such jet-like damsels as before have been mentioned for their light apparel, who with their bravery and white mantles over them seem to be, as the Spaniard saith, "*mosca in leche*", a fly in milk. But the train of the Viceroy, who often goeth to this place, is wonderful stately, which some say is as great as the train of his master the King of Spain.

'At this meeting are carried about many sorts of sweet-meats and papers of comfits to be sold, for to relish a cup of cool water, which is cried about in curious glasses, to cool the blood of those love-hot gallants. But many times these meetings sweetened with conserves and comfits have sour sauce at the end, for jealousy will not suffer a lady to be courted, no nor sometimes to be spoken to, but puts fury into the violent hand to draw a sword or dagger and to stab or murder whom he was jealous of, and when one sword is drawn thousands are presently drawn, some to right the party wounded or murdered; others to defend the party murdering; whose friends will not permit him to be apprehended, but will guard him with drawn swords until they have conveyed him to the sanctuary of some church, from whence the Viceroy his power is not able to take him for a legal trial.

'Many of these sudden skirmishes happened whilst I lived about Mexico . . .'

Perhaps because the Inquisition was not very active in 1625, Gage fails to mention a great stone platform in the western section of this beautiful and fashionable garden. This was the *Quemadero*: it was here that the Inquisition burned heretics. The last great *autos* had been held towards the end of the preceding century: the next would be held during the 1640's. In both cases, the main victims were Marranos, crypto-Jews, who professed Christianity, but secretly practised Judaism. In both cases, economic jealousies would seem to have been as strong among the Creoles as theological zeal: the Jews were, for all the restrictions against them, virtual masters of the commercial system of the Spanish colonies, and were particularly strong in Mexico.

As Gage walked past the Alameda towards the central square, the Plaza Mayor, he would have been entering the section of the fashionable shops. As a general rule, each craft and commodity had a section of its own. Gage tells of

'the beautiful street called *La Plateria* or Goldsmiths Street, where a man's eyes may behold in less than an hour many millions' worth of gold, silver, pearls and jewels. The street of St Austin [San Agustin] is rich and comely, where live all that trade in silks; but one of the longest and broadest is the street called Tacuba, where almost all the shops are of ironmongers, and of such as deal in brass and steel.'

And there were other sections occupied by the workers in feathers, as well as the chandlers, saddle-makers, pastry-cooks and furniture-dealers. Many of the best craftsmen, he notes, are Indians and Christian Chinese: yet, while they handled the most precious materials every day, they somehow never managed to turn them into wealth for themselves.

Mexico was a great trading city; and one of the most colourful descriptions of its traffic is found in Balbuena's *La Grandeza Mexicana,* that curious poem, born between two ages, which has the content of a Chamber of Commerce booster's pamphlet, but is written in the rich allusive style, clotted with sensuous detail and brilliant with glittering images, of the Spanish Baroque. Says Balbuena, rhythmically jingling the heavy gold coins in his sleeve—

> *Riches and opulence are here complete,*
> *Since you'll find treasure here in greater store*
> *Than the cold north can freeze, or the sun heat.*
>
> *Silver extracted from Peruvian ore,*
> *And Chilean gold pour in here every day.*
> *Ternate's cloves, cinnamon of Tidore,*
>
> *Garnets of Ormuz, ransom of Quinsay,*
> *Syrian spikenard, incense of Araby,*
> *Sicilian coral, fine cloth from Cambrai,*
>
> *Diamonds from India, Goan ivory,*
> *Rubies and Scitan emeralds here are seen;*
> *And from Siam comes its grey ebony,*
>
> *With Macao's best, and the best Philippine*
> *Products, and the very best from Spain,*
> *And from both Javas riches peregrine.*

—and the reason for all this wealth is stated elsewhere in the poem—

> *Lucca has not, nor Florence, nor Milan,*
> *Nor any town where bills are promptly paid,*
> *And it's thought good to be a businessman,*
>
> *More noble or more generous a trade,*
> *Merchants more skilled; nor are there anywhere*
> *More goods, more bargains, or more profits made.*

67

In the great square at the centre of the city there had once stood the palaces and council-chambers of Montezuma and his nobles, the massive temple-crowned pyramids of Huit-zilopochtli and Tlaloc, and the great rack where the Aztecs displayed the skulls of their defeated enemies. With its cell-like stone temples, sacred pools, platforms and friezes set in a rigidly geometrical plan, the square had seemed a strange combination of elements: it was at once an Indian ceremonial centre, carried to the highest degree of formal perfection, and something distinctly reminiscent of an ancient Mediterranean temple complex.

Now, however, the Plaza Mayor was the centre of Spanish imperial and civic rule in Mexico. The new square, if not as formally overwhelming as the old, had the heavy authority and sumptuousness of the Baroque. Balbuena, who saw Mexico City not only as a new Carthage, but as a new Rome, and even a new Athens, describes—

> *The portals covered with sculpture—here's largess*
> *Of subtle craft, and here's the richest treasure*
> *Of Corinthian taste and Corinthian tenderness.*
>
> *Flutings and triglyphs subtly vary pleasure,*
> *And, with reliefs of gold in a wide frieze,*
> *Give the whole work proportion and due measure.*

On the east side of the square, very nearly on the site of Montezuma's palace, was the palace of the Viceroy. It presented to the square the stern façade of a fortress, with its clock-tower and its four lesser towers, its regularly spaced loop-holes, its two great gates with their ponderous carved doors, and the Royal Prison ('which is strong of stone work', Gage says) at one end of the edifice. But there were, behind the palace walls, apartments and courtyards of great beauty.

On the northwest corner of the square was the Cathedral of

Mexico, still under construction; its walls had been raised, some domes had been completed, and some wooden roofs laid. On the south side of the square was the municipal palace, where the city council sat. A canal, all that remained of the canoe-basin of Tenochtitlan's great square, ran in front of it. Freight canoes still discharged their cargoes here. On the second storey, there was an open corridor, and at ground level a long arcade running the full length of the building, 'built all with arches on the one side where people may walk dry in time of rain'. Here were the public auctioneers, and petty traders of all kinds: 'there are shops of merchants furnished with all sorts of stuffs and silks, and before them sit women selling all manner of fruits and herbs'. Here the royal officials weighed the silver as it came into the mint (which was just off the other side of the square, beside the Viceroy's Palace), and subtracted from it the tax known as 'The King's Fifth'. Here also was the municipal jail, and the city butcher, who worked under contract to the city.

In its own square, off the Plaza Mayor, was the first European university in America, founded in 1551 by decree of Charles the Fifth. It was not the first college as such, since the Aztecs did have institutions of higher learning, where one could study such subjects as history, good manners, the interpretation of dreams, and astrology. The curriculum of the Royal and Pontifical University of Mexico also included a course in astrology; but one could be excommunicated for casting a horoscope in Mexico, and the art, science, pseudo-science or whatever it may be, was mainly used as an aid in the compilation of almanacs.

This, then, was 'The Imperial, Pontifical and Ever August Mexican Athens' (to quote from an entertainment given at the University in 1683) 'of which city', Gage says, 'a whole volume might be compiled, but that by other authors much hath been written, and I desire not to fill my history with trifles, but only with what is most remarkable in it.'

VIII—*This New World of America*

AT this point, Gage leaves the narrative of his travels to give a description of those parts of North America which Spain claimed as her own. Anticipating what was to be a major preoccupation of the United States, as they expanded from a confederacy on the Atlantic coast into a continental empire, he points out that the Spaniards have declared their intention to conquer all this territory, unless the English 'from Virginia and their other plantations' get there first. He describes the dependence of the Plains' Indians on the buffalo, erring only in what seems to be a belief that they had domesticated them. His report is obviously based on the *Relación* of Casteñeda, a member of the Coronado expedition, which reached Kansas in 1541–2.

> 'The chief riches of this country are their kine, which are to them as we say of our ale to drunkards, meat, drink and cloth and more too: for the hides yield them houses, or at least the coverings of them; their bones, bodkins; their hair, thread; their sinews, ropes; their horns, their maws and their bladders, vessels; their dung, fire; their calves' skins, buckets to draw and keep water; their blood, drink; and their flesh, meat.'

Gage was also one of the first proponents of that theory of Indian origins most popular today; it is supposed by some, he says, that the inhabitants of these northern parts first came over to this New World from Tartary, since 'the west side of America, if it be not continent with Tartary, is yet disjoined but by a small strait'.

70

How had he heard of the strait, whose existence, it is commonly thought, was not suspected until it was discovered by Vitus Bering in 1727?

We do know from the Englishman, Henry Hawks, who travelled in Mexico between 1568 and 1572, that in that period ships were sailing north from Western Mexican ports to discover a western entrance to 'the strait that lies between Newfoundland and Greenland', that is, the Strait of Anian, 'but it was not found'. Earlier, a pilot of the Philippine fleet, Juan de Fuca, who in 1592 discovered the strait which bears his name, had been on such a mission, which had failed.

Gage may also have read, in Hakluyt, an account of the voyage in which Francisco Gali, a pilot on the Macao-Acapulco route, had touched the American coast at a latitude of 37° 30′. Gali had concluded, from the set of the ocean currents, that there was a strait between Asia and America.

Gage's interest in the remote hinterlands was not merely an academic one, nor did it arise from mere curiosity: his references to North America are all part of his campaign to persuade his readers that Britain has a future in America, and could be, if she acted boldly, its sole master. As he says in his preface, addressed to Lord Fairfax, '. . . why should my countrymen the English be debarred from making use of that which God from all beginning no question did ordain for the benefit of mankind?'

His writing is most vivid, though, when he is describing what he has seen himself. There is Chapultepec, 'Grasshopper Hill', near Mexico City, where those viceroys who had died in New Spain were buried, and where those Aztec emperors who had preceded them had also been buried; here 'there is a sumptuous palace built with many fair gardens and devices of water, and ponds of fish, whither the Viceroy and the gentry of Mexico resort for their recreation'.

One of his most picturesque passages describes the convent of La Soledad, also called *el desierto de los leones*. It was built in 1606 for the Discalced Carmelites, as a place of privation, self-torture and prayer. There the monks were expected to live a life as ascetic as that of the Egyptian anchorites of the Thebaid, or for that matter, the Aztec hermit Yappan, who, seeking to please the gods with his austerities, set up his solitary abode on a rock, only to be tempted to lust by the love-goddess Xochiquetzal.

La Soledad was a stately cloister set on a hill, and surrounded by huge rocks, into which cells had been cut. These cells, ten in number, were extremely austere: they were decorated only with religious pictures and images and 'rare devices for mortification, as disciplines of wire, rods of iron, haircloths, and girdles with sharp wire points to girdle about their bare flesh'.

The surroundings were far from austere, though. The cells were set in gardens and orchards some two miles in compass. Springs of water bubbled delightfully among the rocks, and the smell of roses lay heavy in the air. The pain of the hermits was, certainly, aromatic. They only spent a week there, carrying with them a good supply of wines and sweetmeats, and 'as for fruits, the trees about do drop them into their mouths'. The gardens were much frequented by fashionable people, who usually brought along a supply of delicacies for the hermits. It seems there was no sign forbidding this. They also left more expensive gifts. There was a picture in the Church of La Soledad, called our Lady of Carmel, which was surrounded by heaps of 'diamonds, pearls, golden chains and crowns, and gowns of cloth of gold and silver'. 'Before this picture', Gage says, 'did hang in my time twenty lamps of silver, the worst of them being worth a hundred pounds.'

IX—*Gage Runs Away*

GAGE lived in the house at San Jacinto from October to February. As we have seen, he spent much of this time wandering about Mexico City, but he was also careful to inquire into the state of the Philippines, which was, after all, his real destination. During this time he met a friar who had just returned from those islands. The friar's description was not encouraging.

The superiors in the Philippines were cruel, he said, punishing the friars with harsh penances for the smallest and most trifling offences. Some, unable to bear such treatment, had hanged themselves; others, less passive, had been hanged for murdering their superiors.

Yet, though small offences were ruthlessly punished, greater offences seem to have been common. The friar described Manila as a most wicked place, where it was almost impossible to avoid falling into grievous sin. He would have lost his soul forever, had he not escaped by stealth, after many times begging his superior to permit him to return to Spain.

He refused to say more about himself, evading all questions aimed at finding out why he had come away. He would say only that most of the Philippine friars were devils in private, though in public they took pains to appear as gentle as angels and as simple as sheep. Most of them, he said, spent their time studying how to acquire or use the wives and wealth of those under their care. Some of them had been found, in the dim light of morning, dangling on ropes at the gates of their cloisters—murdered and hung up there by the

73

husbands or lovers of their paramours. In some cases, their women had been hung up beside them.

Not all the friars were like this, of course. There were, no doubt, a few saintly men, and there must have been many of sound morals. Yet a number of the friars were, certainly, worldly creatures, and some of them must have more resembled transported convicts than men of God.

The Philippine Church was very rich. Individual traders, who risked their fortunes on a shipment of cargo as one might on a throw of the dice (the eastward Pacific crossing was one of the lengthiest and most dangerous known at the time), were always failing. The Church, however, never failed, not so much because it was built on a rock, but because it acted as a mercantile bank, advancing loans at a rate of twenty to fifty per cent. Since it had huge capital reserves, it could absorb the losses resulting from the bankruptcy of some of its creditors—whose hope of gain lay, with the bones of seamen, at the bottom of the Pacific—and rake in excellent profits from its good debts. It could also take land as security: by now the church held most of the great estates of the islands, which of course yielded them further profit. In Mexico also, it should be said, the religious orders did most of the banking, and held much of the land.

Certainly, much of the money made by these priestly financiers was spent on charitable works. The *obras pias*, such as the *Hermandad de la Misericordia*, performed such necessary tasks as the provision of dowries for poor girls, the relief of the poor, the care of prisoners, the education of orphans, and the support of hospitals and foreign missions. The rapacity of the Philippine church could not be compared with that of the merchants. It is significant, however, that such comparisons must be made, that in its secular actions the Philippine church (the same could be said of the Mexican) offered an example, not of a superior morality, but of a mitigated immorality.

To Gage, who was disillusioned, the friar's tale was the last straw. So he tells us, at least, attributing to himself the best possible motives. However, some moral indignation may have been justified. He came from a family which for more than fifty years had risked everything—possessions, peace of mind, life itself—in the Roman Catholic cause. Had they done all this so that the hermits of La Soledad might wolf sugared fruits in their cells, and Philippine friars, living on usury, might seduce the wives of Manila citizens?

Gage and four of his friends, including Tomás de León, the Irish friar, agreed that they would try to find some way of returning to Spain or remaining in Mexico. The pact had to be kept secret. Had Calvo heard of their intention, he would have imprisoned them in the cloister until the fleet left for the Philippines. They did approach some Mexican friars, to ask what the chances were of their being able to take refuge in some Mexican cloister. The answer was disheartening. There was such a bitter rivalry between the Creole and Spanish friars, the Mexicans said, that they would never be admitted, since nearly all Mexican convents were held by the Creole party. They might find refuge in the province of Oaxaca, where half the friars were from Spain. Their prospects would be even better in Guatemala: there most of the friars were from Spain, and persecuted and slighted those of their brethren who had been born in the country.

It was for this reason that they decided to go to Guatemala—a bold decision, since it was nine hundred miles away, and they possessed neither horses nor money of their own.

At about this time, a certain friar named Peter Borrallo decided, quite on his own, to run away. His destination was the same as theirs—Guatemala. Now Calvo revealed that ruthless efficiency which his appearance, so fat, slovenly and jovial, hardly suggested. He prevailed on the Viceroy to order that the public crier proclaim Borrallo's escape in the market place, and add that any who should harbour Borrallo and refuse to deliver him up, or to hand over any other escaped Philippine friar, should be put in prison, and fined five hundred ducats.

Tomás de León, frightened, withdrew from the pact; and Gage, fearing that he might, in order to worm his way back into Calvo's favour, reveal their plans, pretended to be in agreement with him. The other three friars—Francisco de la Vega, Juan Escudero, and that same Antonio Meléndez who had persuaded Gage to leave Spain in the first place—were completely uncertain as to what they should do next.

At this point, de León drops out of the narrative. We know from other sources that he went to the Philippines with the Dominican mission. Dominican records show that he was sent to Cavite, after having served two years in Manila. There is no further record of him.

When de León was not within earshot, however, Gage continued to encourage his companions. You see, he said, what Calvo thinks of his friars, to have their escape cried aloud in the market-place, as if they were runaway slaves. Why should they consent to be shipped like cattle to the other end of the world? In any case, he had heard privately that Borrallo had been seen, far from Mexico City, travelling alone towards Guatemala. Could they not escape as well as he?

It was agreed, therefore, that the night before the mission left for Acapulco to join the Manila fleet, the four would slip away from San Jacinto. Two by two, they would go to Mexico City, where they would have horses waiting. They would travel by night, until they had gone some twenty or thirty leagues. Calvo would not delay to search for them, since, if he and his company missed the sailing of the fleet, he would have to wait another year in Mexico. It is true that the Viceroy had appointed officers to watch the chief roads out of the city until the Dominican mission left for Acapulco, but they could easily avoid capture by leaving Mexico City through the side streets.

A 'friend', whom Gage does not name, had offered to guide them out of the city by an obscure route: he had also prepared a map for them. It was decided they would leave through the suburb of Guadelupe, thus giving the impression that they were heading away from Guatemala, rather than towards it.

After the four had purchased their horses, they pooled their money in a common purse. They found they had only twenty ducats, or slightly more than twenty English shillings in purchasing power. Yet they resolved to go on, relying, as Gage tells us, more on the Providence of God than on any earthly means.

They left around the middle of February—just in time— at ten in the evening, and they travelled comfortably till dawn. At a little Indian town, they ordered a turkey and a capon, and breakfasted with their mysterious friend and guide before he returned to Mexico. They slept for the rest of that day, and set out again at nightfall. When they reached the town of Atlixco, they stayed at a farm, set well back from the main highway.

They were already beginning to feel more at ease. They were not, after all, sleeping in barns, and nobody seemed to regard them with suspicion. The richer farmers and yeomen they encountered were honoured by their visit, and entertained them sumptuously; they had great respect for priests and friars, and nobody was presumptuous enough to ask them where they had come from, and where they were going. Usually, when they left, their hosts gave them money. By the time they were forty leagues out of Mexico, the common purse had doubled its weight; they had forty ducats.

In the valley of San Pablo, not far from Atlixco, they came upon a farm owned by a countryman of Meléndez', who had come from the friar's own town of Segovia, in Spain. For the sake of Segovia, this yeoman kept them with him three days and nights. He was well off: his sideboard was full of silver bowls and cups; his table was set with plates rather than wooden trenchers, and his daughters were trained in music. The valleys around Mexico were full of such farmers, who, like their North American counterparts of more than a century later, had vastly improved their condition in the New World.

From San Pablo—they were following, deliberately, a very circuitous route—they came to Taxco, then a small hill-built town of some five hundred inhabitants. Then, after moving through some smaller towns, they came to Oaxaca— a small city of about two thousand inhabitants, Gage tells us, yet fair and pleasant to the eye. It was the centre of a rich cattle and sheep district; the religious houses were notably well off, and the town was particularly well-known for its chocolate, which was put up in boxes by the nuns of one of the convents, and sent all over Mexico, and, indeed, as far as Spain.

Here they were informed of a law most convenient to religious travellers. The Indians were obliged to provide, free of charge, to all friars passing through their towns, a change of horses and food to carry them farther on their journey, provided only that the friars stayed no longer than twenty-four hours. At their departure, the cost of this service was to be entered in the town book, which at the end of the year was to be shown to the Spanish justice under whose jurisdiction they were. The amount would then be repaid from the proceeds of the sale of the wheat or maize grown on the common plot of the town.

Here, also, they encountered for the first time unmistakable evidence of the virulence of the hatred which existed between the native-born Spanish-Mexicans and the *gachupines* from Spain.

There had been in Oaxaca a grave old friar from Spain. He was a Master of Divinity, and noted through the entire valley for his learning. Yet, because he was from Spain, his reputation had only aroused the hatred and envy of the Mexican friars. While he was alive, they had been unable to discover anything against him. But after he had died, they had discovered in his chamber a coffer containing a sum of money which he had not, while he was alive, declared to his superior. The old man violated his vows, they cried; he has died in sin. So they buried the learned old man, and his reputation with him, in the middle of the garden of the convent. The entire valley cried out against the friars for this act; yet they responded by saying that since he had committed a sin worthy of excommunication, and had neither confessed it nor made restitution, he had died excommunicate.

Yet all of them, Gage tells us, were guilty of this same sin of making their vocations pay; some of those who condemned the old man must have sinned much more greatly than he. It was because he was from Spain that they did this, he says, so great a hatred did they bear to the *gachupines*, who both by law and custom occupied all the high positions in the country.

The four renegade friars decided they would not stay in

Oaxaca. What security would they have, three Spaniards and a Hispanified Englishman, in this snake-pit of intrigue? They might at any moment be condemned and turned over to the secular authorities, to be shipped, under heavy censure, to Manila.

After three days, they decided to move on. Their destination was Chiapa, which was three hundred miles from Oaxaca, and part of the Dominican province of Guatemala. They thus began to travel towards Tehuantepec, which was on the shore of the Pacific, and stood at the beginning of the Guatemala road.

X—*From Tehuantepec to Chiapa*

TEHUANTEPEC was a fishing town: you could meet on the roads long trains of fifty to a hundred mules, their backs piled high with salt fish for Oaxaca, Puebla and Mexico. It was also a small and prosperous port whose merchants, some of whom were very rich, traded with Peru and other colonies to the south.

When Gage and his companions arrived in Tehuantepec, they found they had the choice of two roads: they could take the coast road to Guatemala, which passed through the provinces of Soconusco and Suchitepequéz, or the mountain road which would take them to Chiapa, through wild and rocky country. They chose the latter.

As they rode away from Tehuantepec, the land became drier. After they had passed through the little town of Ixtepec, they found themselves in a desert—the beach sand to their right, and to their left an earth almost as dry as sand. The water courses were only dry cracks in the earth; the

St Jondelus.

Boutron

S^{anlle} croux

Vera Cruz and San Juan de Ulloa, as seen by Champlain in 1599

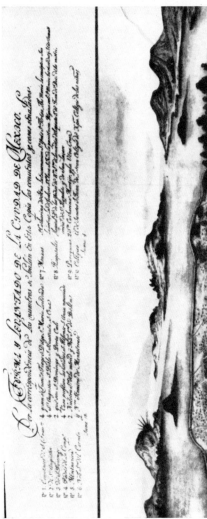

Mexico City in 1628—the celebrated *Vista* of Juan Gomez de Trasmonte

green of the scanty vegetation—low brush, organ cactus, and one or two trees deserving of the name—was dull with dust.

They travelled through the desert for two days. There were no towns or houses; there were only, at wide-spaced intervals, thatched lodges for the shelter of travellers. One night they did not reach a lodge before nightfall, and slept on the ground. The plain was open to the sea: the wind blew continually, and with such strength that they were almost blown off their horses. Occasionally they would see a few cattle or some wild horses, but they encountered no human beings.

On the second day, Gage became separated from his three companions. They had ridden on ahead, hastening to reach the next town before night-fall. Gage, who was having trouble with his horse, was following them at some distance. The horse grew more and more obstinate: finally it refused to move any farther. Gage dismounted, intending to lead the animal to the town, which he knew could not be far; but at this point, the horse lay down.

Gage now had two alternatives, neither of them attractive. He could stay where he was. He could set out for the town on foot, leaving his horse to browse; but in that great flat plain, with not a hedge, tree or shrub to mark the place or hide the saddle, he might never find him again.

He decided therefore to pass the night where he was, hoping his friends would send from the town to see what had become of him. He unsaddled his horse, and lay down to rest, using the saddle as a pillow.

He lay there for about an hour, watching his horse feeding on the short, withered grass. He was about to go to sleep, when he heard a fearful howling, barking and crying. At first, when it was a good distance away, it sounded to him like the noise made by a pack of dogs. When it came closer, he realized that it was not a sound such as dogs make; there seemed to be human shrieks intermingled with the animal noises. Now it was an age still acutely aware of the supernatural, and Gage, so many times a renegade, had an uneasy conscience—

'Sometimes I thought of witches, sometimes of devils, sometimes of Indians turned into the shape of beasts (which amongst some hath been used), sometimes of wild and savage beasts, and from all these thoughts I promised myself nothing but sure death, for the which I prepared myself, recommending my soul to the Lord, whilst I expected my body should be a prey to cruel and merciless beasts; or some instruments of that roaring lion who in the Apostle goeth about seeking whom he may devour. I thought I could not anyways prevail by flying or running away, but rather might that way run myself into the jaws of death; to hide there was no place, to lie still I thought was safest, for if they were wild beasts, they might follow their course another way from me, and so I might escape. Which truly proved my safest course, for while I lay sweating and panting, judging every cry, every howling and shrieking an alarm to my death, being in this agony and fearful conflict till about midnight, on a sudden the noise ceased, sleep (though but the shadow of death) seized upon my wearied body, and forsook me not, till the morning's glorious lamp, shining before my slumbering eyes and driving away death's shadow, greeted me with life and safety.'

Blinking, and looking about him, Gage saw that his horse was still there. He expressed in prayer his gratitude for having lived through the terrors of the night; then he rose from his knees, and saddled his horse. He wanted nothing so keenly as to be out of that place.

After he had ridden rather more than a mile, he came upon a brook, where two roads met. One led through the desert; he could gaze down that road for five or six miles, and see neither a house nor a tree. The other led to a small grove of trees, some two or three miles off. Here, he thought, there might be a town.

After he had ridden a quarter of a mile farther, his horse grew stubborn again. Gage dismounted, and led his mount by the bridle, until, as he approached a thatched house by the side of the road, he saw an Indian riding towards him. Gage asked him the distance to the nearest town. The Indian answered that it was just beyond the trees, but Gage would not see it until he was almost on top of it. Encouraged now, Gage mounted again, and spurred and prodded his horse until they came to the trees. There the horse stopped, and refused to go any farther. Gage tethered him, unsaddled him, hid the saddle under some low shrubs, and entered the town on foot. Soon he found his friends. Not in the least suspecting that he had spent the night in the desert, they had sent to another town to inquire if he had been seen there.

When Gage told the Indians of the fearful noise which had kept him awake and in terror, they answered that they heard such howlings every night: they were only the cries of 'wolves and tigers', which came to steal their livestock. They commonly frightened them away with sticks or holloaings.

The next morning, after Gage had sold his horse and hired another, the four set out again.

They moved from town to town, encountering no difficulty and braving no danger or deprivation. They were usually entertained by the Indians or by the local priest, and they seem to have been entertained well. Eventually they came to Tapanatepec, at which point the land began to rise into the Chiapa Highlands. These were frightening mountains, and they had heard terrible stories of them. It was said they were the most dangerous mountains in Mexico, and that, near the peaks, one had to pass through some very high narrow passes, through which the winds from the Pacific, whose wrinkled skin one could see far below, rushed with such fury that they had on occasion blown laden mules, and horses with their riders, to a bloody death. Nevertheless, the friars decided they would face the passes, provided only the winds were not too boisterous on the day they began the climb.

Tapanatepec, nestling at the foot of this range, was an attractive and prosperous village. The climate was spring-like. The wealthier Indians had from a thousand to four thousand cattle on their estates; the sea, and the little river which ran by the town, were rich in fish; and the many streams, falling from the mountains which overshadowed the town, watered fertile market-gardens and orchards of orange, lemon and fig.

They set off up the mountain of Macuilapa in the afternoon. They were on muleback. Two Indians were guiding them, and in their saddlebags were fruit, fried fish and a cold roasted capon. They had been told that they need only climb some twenty-one English miles through the dangerous part of the mountains. After that, it was said, they would reach a country of wealthy ranchers, chief among whom was one Don Juan de Toledo, who would make them welcome. That night, they camped in a green meadow tucked away between two of Macuilapa's rocky ribs. They ate their capon and most of their fish, leaving only a few scraps for breakfast; then they turned in, lulled to sleep by the trickling of the brooks down the side of the mountain.

In the morning, they continued the climb. They had gone barely a mile when the wind began to whistle about them, and they were afraid. They thought of returning, though they were only halfway up the mountain; but the guides encouraged them, saying that there was a spring about a mile farther on, and that there they would find a lodge, built under the trees to shelter travellers.

They took refuge in the lodge that night, the more grateful for the trees around it, which broke the incessant pressure of the wind. They began to wish they had kept some of the provisions they had so thoughtlessly eaten, even some of the refuse they had thrown away: they would gladly have sucked the head of a fried fish, if they had thought to save one. At last, while they were searching among the trees around the lodge, they found a lemon tree. It had gone wild, and its fruit was small, green and very sour. Nevertheless, they plucked it greedily, and the flesh of green lemons, its acidity somewhat diluted by water from the spring, was their supper that night.

The next morning, they had more green lemons for breakfast. In the early afternoon, one of the friars, his eye sharpened by hunger, noticed that the Indians carried with them, as travel rations, little bags of powdered maize. They bought half a bag, for which the Indians charged them the equivalent of two shillings. In Tapanatepec, it might have been worth a penny.

Over Tuesday and Wednesday they waited for the wind to die down, living, for those two days, on green lemons, maize powder and water. On Thursday morning, the wind was as strong as ever; but they resolved to try the pass, since they were weakening, and might well starve to death, or die of sickness, on the side of Macuilapa.

Upon the bark of a giant tree they wrote their names, and the number of days they had been in the lodge without food. Then, feebly mounting their mules, they began to climb once again.

The way grew narrow and steep, and they alighted from their mules, thinking themselves safer on their own feet. Every hour, as they climbed, the wind grew stronger.

When they reached the top of the Macuilapa pass, they found it to be a bare rocky place with not a tree, not even a shrub to hinder the fury of the wind. They were on a ridge which was, in places, no more than forty-five inches wide. On one side they could see, horribly far below, the dazzling expanse of the Pacific; on the other side, they could see a rocky descent of, Gage says, six or seven miles. They crept forward on this narrow ridge of rock: the Indians led the way, and the friars, weakened by hunger, not daring to look to either side but staring at the ground, followed them on all fours. Their habits whipping in the wind, they inched forward, placing their hands and knees in the prints left by the hooves of the mules.

After what seemed hours of this frightful progress, the track widened, and they approached a clump of trees. They lay down, gasping with exhaustion, cursing themselves for their folly in seeking to travel the direct route over Macuilapa.

In any case, they were out of danger now, and so relieved were they that they moved, in spite of their weakness, with fair speed. Soon they arrived at the ranch of Don Juan de Toledo, who gave them some warm broth; but they were so weak that they promptly vomited it up again. After taking many small sips of broth and wine, they found they had recovered enough strength by nightfall to be able to cope with supper.

They rested two days with Don Juan de Toledo, and on the third they set out for Chiapa de Corzo.

XI—*Chiapa*

CHIAPA DE CORZO was, in Gage's time, called Chiapa of the Indians, to distinguish it from Chiapa proper, which was inhabited by Spaniards. It overlooked the Grivalja, which river, Gage states, is as broad as the Thames at London. It was a prosperous town: its Indian governor, Don Felipe de Guzmán, kept a dozen horses in his stable, was allowed to wear a rapier and dagger, and had not only sued the governor of Spanish Chiapa over some privileges of the town, but had actually won his case. 'No town', says Gage, 'hath so many dons in it of Indian blood as this.'

Here Gage and his companions were well-received by their fellow Dominicans. The Provincial and Chief Superior, they were told, would be glad to see them: the Creole and native friars, striving to control the convent as their countrymen controlled those in Mexico City, had the few Spaniards at a disadvantage. There was no doubt the newcomers would be taken in.

Here, by what would at first seem to be an amazing coincidence, they met Peter Borrallo, whose earlier escape had encouraged them in their own. In fact, though, there was nothing amazing about their meeting: there were not that many roads from Mexico into Guatemala, and Borrallo, as a Dominican, would be staying, as he travelled, in Dominican houses. Borrallo told them they had nothing to fear: he had been treated very kindly, and, in any case, Calvo and his train had already sailed from Acapulco. True, Calvo had loosed a Parthian shot on his departure: he had written a harsh letter to the Superior of Chiapa and Guatemala, in which he had asked him to order his provincials not to shelter Borrallo, Gage and their friends, but to send them back to Mexico under guard, so that they might be held there till the sailing of the next fleet for the Philippines. However, the Provincial in Chiapa, Father Pedro Álvarez, had not paid the slightest attention to the letter, and Borrallo had not been molested.

Álvarez was now in a little town just a day's journey from

87

Chiapa de Corzo. They decided to present themselves to him there. Since they could not hope to find refuge in any other part of America, they would ask him if they could either stay in Chiapa and Guatemala, or return to Spain.

When they arrived, they found him walking through one of the shady byways of the town. He entertained them that night, speaking to them in a gentle and fatherly manner; and before they went to bed, he washed their feet with his own hands.

That first day, he said almost nothing on the subject of their continuing under his jurisdiction. The next day, though, he produced the letter from Calvo, and read it aloud, adding his own severe comments—that they had sinned in forsaking their call to the Philippines, that they might have saved many Filipinos who would now be damned for having been denied missionaries of their ability, that they had betrayed and disappointed their King, at whose expense they had travelled from Spain, and, finally, that they must consider themselves his prisoners, and that he was empowered to ship them back to continue their journey, as Calvo had demanded. In the meantime, since he did not know what to do with them, they should take their ease in the grounds of the house he was now staying in. After dinner he would have an answer to a letter he had written to the Superior in Chiapa.

He then showed them into the garden.

They paced back and forth in the shade of the orange trees. Their spirits were low. Álvarez had talked in high terms: he had used words like 'souls', 'king', and 'imprisonment'. Could they doubt that they would be shipped back to Mexico? Each friar dwelt on his own black thoughts: Gage thought to himself that he would never see England again, but would moulder his life away in the Far East; Meléndez wished he were anywhere but where he was, even on the high pass at Macuilapa; another wished he had never flown from Mexico, but had gone quietly to the Philippines with Calvo. What could they do? One proposed that they try to escape from Álvarez, but another answered that no matter where they fled in Guatemala, they would be discovered. Even if they were sent to Mexico, he said, it would be easier to escape from the party escorting them there than from their present semi-captivity. Gage himself added some cogent arguments—that the Provincial had received them gently, even washing their feet as Our Lord had washed the feet of His apostles, that he could hardly condemn them for having come all that way to seek his protection and to offer them-selves 'as fellow-labourers in that harvest of souls belonging to his charge', that he wanted Spaniards to counter-balance the Creole faction, that since he had accepted Borrallo he would hardly reject them, and, finally, that even if he did not accept them there, he would probably send them to Spain rather than to Mexico.

While they were thus exchanging worries and doubtful solutions, old Álvarez was watching them from his window. He sent another friar out to them, who asked them, as he approached, why they were so melancholy. The Provincial, said the friar, had observed their sadness. They should know that their melancholy was unjustified. The Provincial wished them well, and needed them; and since they had thrust themselves on his mercy, he would hardly be so cruel as to do to them what martial law forbade even a soldier to do to his defeated enemy.

The Creole faction, he added, had harshly censured the Provincial for sheltering Peter Borrallo: there could be no doubt that they would complain even more bitterly at the arrival of four more *gachupines*, and renegades at that. The Provincial could hardly be blamed for wanting to consider the matter well. For their parts, they should carry themselves with such discretion that those who condemned even the best of the Provincial's actions would have no arguments against their staying. In any case, if for some reason they could not be admitted into Chiapa or Guatemala, they would not be sent back to Mexico, but would be allowed to return to Spain.

Now the bell rang for dinner; and Gage, who could always be cheered by the prospect of a good meal, went in with a light heart, followed by his three renegade companions and the Chiapan friar.

After dinner, the Provincial invited them to a game of backgammon. Gage, in spite of his strictures against clerical gambling, gladly assented. In any case, they were not to play for money. Since I know you are poor, the Provincial said, we will play on the following terms: if I win you owe me five *Pater Nosters* and five *Ave Marias*, but if I lose I will be compelled to admit you into the province.

The game began. The three young friars, taking turns one after the other, were too much, as it seemed, for the old man; but Gage, who was more acute, sensed that he was willing to lose 'that his very losses might speak unto us what through policy and discretion he would not utter with words'. Indeed, a messenger arrived as soon as they had finished the game, bearing a letter from the Prior in Chiapa. The Prior, replying to the Provincial, expressed great joy at the arrival of the four friars, and begged that they be sent him immediately, to be his guests. For, he said, their case had been his some two years before. He too had been in a company of Philippine missionaries; he too had deserted them, and fled to Guatemala. He was glad to welcome them; and, he added significantly, he hoped they would join with him against those who spited him—that is, the Creoles.

Álvarez declared himself to be very pleased by this letter; as to the game of backgammon, he would now pay them what he had lost. The next day, he said, he would send them to Chiapa. They must stay there until he was ready to send them to other parts of the province, where they would learn the languages of the Indians, and preach to them.

The four friars went into the garden once again, but this time they were in a happier frame of mind. Indeed, they began to play there like children, laughing, joking and pelting each other with oranges and lemons, their particular target being the friar (probably the sub-deacon Juan Escudero) who had wished he were on the boat with Calvo. 'We strived to beat him out of the garden by force of orange and lemon bullets; which sport we continued the more willingly, because we perceived the good Provincial stood behind a lattice in a balcony beholding us, and rejoicing to see our hearts so light and merry.'

After supper, they had another game of backgammon with the Provincial; and now the stakes were different: if they won, he would give them a box of chocolate, and if they lost they would be his prisoners. This time the Provincial won: in fact he was such a cunning player that he could win any game he chose. I am sorry you have lost, he said, but I hope that you will never be prisoners to a worse enemy than I am. To console them, he would give each of them a box of chocolate, and they must drink some of that chocolate, for his sake and for their own comfort, when they were most disheartened by their imprisonment. They thought this was nothing more than a kindly joke, and they paid little attention to it.

In the morning, there were mules waiting for them, and an escort of twelve mounted Indians. They said goodbye to the Provincial, and set out for the town of San Felipe.

On their arrival, they met the Prior of Chiapa, Father Juan Batista. He took them to breakfast, with the rather disquieting remark that they would do well to eat heartily, since their dinner that night would be the meanest thing they had ever eaten. He further admonished them to enjoy their liberty, since it would not last long. They had been told they were out of danger, so they did not know how to take the remark. Was it a threat, or a mock-sadistic joke? After breakfast, the Indians of San Felipe entertained them with horse-races and *juego de cañas*, a sort of jousting. But they could not stay long, since they were expected to arrive in the cloister of Chiapa, two miles away, before noon. Their mules were brought to them; and the Indian singers and trumpeters (as a conquered and now servile race, the Indians were expected to do this kind of thing) escorted them 'with dances, music and ringing of bells' a half-mile out of town. Then the Prior dismissed them.

When they were within half a mile of Chiapa, the Prior ordered them to halt. He took a paper from his pocket, an order from the Provincial; and he proceeded to read it aloud in a formal manner. It said:

> '. . . That whereas we had forsaken our lawful Superior Calvo in the way to the Philippines, and without his licence had come unto the province of Chiapa, he could not in conscience but inflict some punishment upon us before he did enable us to abide there as members under him; therefore he did strictly command the Prior of Chiapa that as soon as we should enter into his cloister, he would shut us up by two and two in our chambers, as in prisons, for three days, not suffering us to go out to any place, save only to the public place of refection (called refectory) where all the friars met together to dine and sup, where at noon time we were to present ourselves before all the cloisters sitting upon the bare ground, and there to receive

no other dinner but only bread and water; but at supper we might have in our chambers, or prisons, what the Prior would be pleased to allow us.'

The Prior, noting their sadness, added a comforting postscript of his own. He assured them that after their short imprisonment, they would be honoured and preferred, and that, though their dinners should be of bread and water, he would afterwards send them a good supper to their chambers, which would more than sustain them for the next twenty-four hours. Thus encouraged, they proceeded to the cloister of Chiapa, where most of the friars greeted them cordially. But the faces of some, they noticed, were surly and frowning.

Almost as soon as they had been conducted to their chambers, the dinner-bell rang. It summoned them, however, to something less than a dinner. They went down to the refectory, and after grace had been said and the friars of the convent had seated themselves, the four 'Philippine Jonahs' (as some of the Creoles had decided to call them) sat cross-legged, 'like tailors', on the bare ground. Each of them received a loaf of bread and a pot of water.

A Creole friar was sitting beside them, doing penance for having written some love-letters to a nun. He did not consider his sin as great as theirs, and, perhaps to disguise his own sense of shame, he looked at them with an expression of exaggerated disgust. Gage (one of his redeeming features appears to have been a certain mischievous quality) chose a place as near to him as possible. The friar, under his breath, muttered 'disobedient Philippine Jonahs', to which Gage responded, in a soft and friendly tone, with two hexameters, invented on the spot—

> *Si monialis amor te turpia scribere fecit,*
> *Ecce tibi frigidae praebent medicamina lymphae.*

(If love for a nun made thee write scandalous things,
Lo, cold waters furnish thee their medicines.)

The friar, much offended, tried to remove himself (his penance forbade him to rise until dinner was over) by wriggling his elbows and shoulders and inching away on his buttocks; but Gage followed him in the same manner, adding—

Solamen misero est socios retinere pañetes.

(It is a solace to one in trouble to recall that a trouser leg always keeps its partner.)

The friar thought Gage was after the loaf which he was busily gnawing, so he hugged it closer to his breast. But the word *pañetes*—a Spanish word for a kind of trousers—so amused him in its Latin context that he began to laugh, and almost choked on his mouthful of bread.

Having eaten their meagre dinners, the four Jonahs returned to their cells. Shortly after, the Spanish friars began to crowd in, bringing them conserves, sweetmeats and other dainties, and begging them to tell their story. The Prior himself supped with them that night, and the food was excellent.

Even their being deprived of liberty over the next three days was a favour rather than a punishment, since it enabled them to rest after their long journey.

When they were set at liberty, the Provincial and Prior began to consider what might be done with them. Eventually two of the friars (Gage does not give their names) were sent into the country to learn the local Indian language. Gage and the remaining friar asked to be sent to the famous university at Guatemala, to study philosophy and divinity. This was agreed to, but it was decided that nothing could be done until Michaelmas, the beginning of the 'school year'.

In the meantime, the Provincial had heard of Gage's

94

extempore Latin verses. A teacher of grammar and syntax was needed in the school attached to the Chiapa cloister, and since English scholars were thought at that time to be better Latinists than the Spaniards, he appointed Gage to this position, granting him an allowance for books and other necessaries.

Gage remained in this position from April to the end of September.

Chiapa, a poor town of no more than four hundred Spanish householders, was served by a cathedral, and by two cloisters —one of Dominicans, the other of Franciscans. In additions there were about a hundred Indian families in a suburb, with a chapel of their own. The fact that there were no Jesuits in Chiapa, says Gage, is a proof of the poverty of most of the inhabitants, and the miserliness of the few rich ones, Jesuits usually attach themselves, 'like horse-leeches', to the wealthy and gallant. In Chiapa, however, the merchant: hugged their wealth to their bosoms, and the gentlemen were a wretched, threadbare, penurious lot.

> 'The gentlemen of Chiapa are a by-word all about that country, signifying great dons (*dones*, gifts or abilities, I should say), great birth, fantastic pride, joined with simplicity, ignorance, misery and penury. These gentlemen will say they descend from some duke's house in Spain, and immediately from the first Conquerors; yet in carriage they are but clowns, in wit, abilities, parts and discourse as shallow-brained as a low brook, whose waters are scarce able to leap over a pebble stone, any small reason soon tries and tires their weak brain, which is easily at a stand when sense is propounded, and slides on speedily when nonsense carrieth the stream . . . As presumptuous they are and arrogant as if the noblest blood in the Court of Madrid ran through their veins. It is a common thing amongst them to make a dinner only with a dish of *frijoles* in black broth, boiled with pepper and

garlic, saying it is the most nourishing meat in all the Indies; and after this so stately a dinner they will be sure to come out to the street-door of their houses to see and be seen, and there for half an hour will they stand shaking off the crumbs of bread from their clothes, bands (but especially from their ruffs when they used them), and from their mustachios. And with their tooth-pickers they will stand picking their teeth, as if some small partridge bone stuck in them; nay, if a friend pass by at that time, they will be sure to find out some crumb or other in the mustachio (as if on purpose the crumbs of the table had been shaken upon their beards, that the loss of them might be a gaining of credit for great housekeeping) and they will be sure to vent out some non-truth, as to say: *O Señor, que linda perdiz he comida hoy*, "O Sir, what a dainty part-ridge have I eat today", whereas they pick out nothing from their teeth but the black husk of a dry *frijol* or Turkey bean.'

As for the women, they were equally bizarre creatures. So weak and squeamish were their stomachs, they claimed, that they must always be drinking chocolate, even in church. Indeed, during a solemn high Mass, the cathedral was more like a coffee-house than a church: there were always a number of maids bustling about, bringing their mistresses chocolate or taking away the empty cups.

Just before Gage had come, the Bishop had forbidden the drinking of chocolate in the cathedral, and when this had had no effect, he had fixed a notice to the church door, to the effect that all who drank chocolate or nibbled sweetmeats during Mass would be excommunicated. From this moment, the women of the town conceived a great hatred for the Bishop. Some of them missed no opportunity to insult or slight him; others simply continued to drink chocolate in church as they had before. One day, when the canons and other priests attempted to take away the cups of chocolate from the maidservants, the women's husbands drew their swords against them, and they had to retire to avoid bloodshed.

Finally it became clear to the ladies that the Bishop would neither be persuaded nor coerced, and they decided to leave the Cathedral. Most of the fashionable people of the city resorted to the cloister churches. There was a degree of rivalry between the secular priests and the religious orders; and the result was that the cloisters grew rich, and the Cathedral, deprived of its offerings, grew poor.

At this, the Bishop posted another order, demanding that everybody in the city resort to the Cathedral. The response of the women was to stay at home. For a month the ladies of Chiapa did not attend church at all.

Now the Bishop developed an inexplicable sickness. Eventually he went to the Dominican cloister, where his friend, the Prior, took care of him. The doctors came; and it was their unanimous opinion that the Bishop had been poisoned. None of them could save him, and he weakened from day to day. He died, praying that those who had caused his death might be forgiven, and that the sacrifice of his life, offered for the honour of God and in the zeal of His house, would prove acceptable in His sight.

The common talk around town was that the Bishop had been poisoned by a gentlewoman who was somewhat too familiar with one of the Bishop's pages. It was said that she had, through this page, sent him a cup of poisoned chocolate. To the ladies, this seemed no more than poetic justice. Gage himself heard this gentlewoman say that few grieved for the Bishop, least of all the ladies, and that she supposed, since he was such an enemy of chocolate in church, that the chocolate he had drunk at home had not agreed with him either.

Afterwards it became a proverb in that country: 'Beware of the chocolate of Chiapa.'

The story is not so fantastic as it appears. The air of Chiapa was poisonous with rivalry and conflict. The Creoles resented the economic power of the church, even while they supported it; but they resented even more those few priests, zealous for justice, who condemned them for their greed, materialism, and ruthless exploitation of the Indians.

Fray Bartolomé de las Casas, that zealous if unbalanced fighter against oppression, had been Bishop of Ciudad Real de Chiapa in the preceding century. He had found the experience discouraging. In his time (the middle 1500's), he could remember only two governors and one bishop who had 'feared God, or obeyed God's laws or the King's': the colonists, for their part, replied that Las Casas was an 'ill-humoured, unquiet, importunate, noisy, trouble-making fellow', a complaint which has been made against other socially-conscious priests, in other places.

Perhaps the key to this story is to be found in Gage's statement that the Bishop was, though somewhat covetous, 'very zealous to reform whatever abuses [were] committed in the church'. Moral zeal was a rather dangerous virtue in seventeenth-century Mexico; it was particularly dangerous in Chiapa.

In fact Gage himself, by rejecting the advances of the suspected poisoner, placed himself in some danger.

'The women of this city are somewhat light in their carriage, and have learned from the Devil many enticing lessons and baits to draw poor souls to sin and damnation; and if they cannot have their wills, they will surely work revenge either by chocolate or conserves, or some fair present, which shall surely carry death along with it. The gentlewoman that was suspected (nay was questioned for the death of the Bishop) had often used to send me boxes of chocolate or conserves, which I willingly received from her, judging it to be a kind of gratuity for the pains I took in teaching her son Latin. She was of a very merry and pleasant disposition, which I thought might consist without sin, until one day she sent unto me a very fair plantain wrapped up in a handkerchief, buried in sweet jasmins and roses; when I untied the handkerchief, I thought among the flowers I should find some rich token, or some pieces of eight, but finding nothing but a plantain, I wondered, and looking further upon it, I found worked upon it with a knife the fashion of a heart with two of blind Cupid's arrows sticking in it, discovering unto my heart the poisoned heart and thoughts of the poisoner that sent it. I thought it a good warning to be wary and cautious of receiving more presents and chocolate from such hands, and so returned unto her again her plantain with this short rhyme cut out with a knife upon the skin, *Fruta tan fria, amor no cria*, as much as to say, fruit so cold, takes no hold. This answer and resolution of mine was soon spread over that little city, which made my gentlewoman outrageous, which presently she showed by taking away her son from school, and in many meetings threatening to play me a Chiapaneca trick. But I remembered the Bishop's chocolate and so was wary, and stayed not long after in that poisoning and wicked city, which truly deserves no better relation than what I have given of the simple dons, and the chocolate-confectioning donnas.'

XII—*Chimaltenango*

By the end of September (Michaelmas), it was time for Gage to leave Chiapa for the City of Guatemala.

On the way, he had an alarming experience.

In the town of Chimaltenango, in the highlands of Guatemala, he stayed with a certain Dominican friar named

Alonso Hidalgo, an old man who wore spectacles, and had false teeth of lead, both facts worthy of note in those days. Hidalgo was a bitter man, and an ambitious one. Though he had been born in Spain, he had been raised in Guatemala, and it was in Guatemala that he had taken his vows. He had become more Creole than the Creoles; he hated all who came from Spain, and it was his hope that, with the help of his Creole friends, he would become the Provincial. It was because of his native sourness and suspicion, Gage tells us, that he conceived the peculiar notion that Gage was a spy.

On the way to Chimaltenango, on a dangerous mountain track, Gage's mule had thrown him, and both mule and rider had rolled down the almost perpendicular incline. They would have been smashed on the rocks below, had not a shrub broken Gage's fall, and a tree the mule's. The Indians who were escorting him at once cried out *milagro, milagro,* 'miracle, miracle', and *santo, santo,* 'a saint, a saint'. The word preceded them along the road, and when they reached the next town, Gage was welcomed as a man of unusual sanctity. His impatience at being so honoured was taken by the Indians as merely an instance of his humility, and therefore a confirming indication of his holiness. In the town of Joyabaj, he sat in a special chair in the choir of the church: all the Indians of the town came up to render him homage, and to offer gifts of coins, honey, eggs, fruit and fowls. Gage, most embarrassed, asked the friar who ministered to them, one Juan Vidal, to disabuse the natives of this notion, something he could not do himself, since he possessed only a partial knowledge of their language. Instead, the friar addressed the natives to thank them 'for the great love they had shown unto an ambassador of God'.

Later, when Vidal totted up the take, he expressed his

gratitude to Gage, who had filled his larder for many days. Gage's cash reward for almost having broken his neck was forty reals, and Vidal gave him twenty more for the food offerings, which were, Gage tells us with some resentment, worth at least forty. He also gave Gage some friendly advice: such simple-minded errors should not be discouraged; for so long as the Indians thought the friars to be on the very brink of divinity, so long would they obey them in everything, so that their persons and fortunes could be commanded at pleasure.

Gage's reputation as a holy man had also preceded him to Chimaltenango; but Hidalgo's reception was cold. He simply refused to believe that anybody who came from Spain could be a saint, much less a man from England, that country of heretics. Rather, he thought, Gage must be a spy, come to view the riches of Guatemala, and afterwards betray them to England. There were, after all, many treasures in the Dominican cloister at Chimaltenango, notably a picture of Our Lady and an expensive lamp, and Gage would, he had no doubt, inquire into these. To this Gage answered in the only possible way, with a joke. He would first take notice, he said, of the pictures, tapestries, and rich cabinets in Hidalgo's chamber. He might even tell the English about Hidalgo's teeth, in which case they would make him drink very hot soup, to see if they were indeed lead, and not silver. This would certainly cause them to melt, and run down his throat. After a number of such agreeable pleasantries, 'he perceived I jeered him, and so he let me alone'.

He did not stay for supper under the roof of the suspicious old man, but set out again for Guatemala City. There, he said, he would have a light supper in the Dominican friary.

What Gage did not know then—probably he never did know—was that an order respecting him had been sent out in that year of 1627 to the Guatemalan Provincial. It had come from the Master General of the Dominican Order in Rome. It had ordered that Gage be sent to Spain 'as our English brethren need his services for the propagation of the faith'. But the order had not reached its destination. Gage was to be in the Dominican friary at Guatemala for the next three years.

XIII—*The Highland Mayans*

On his way to Guatemala, Gage would be passing through the country of the highland Mayans—the Quichés, Zutuhils, Cakchiquels, Pokomans and other related peoples. Since these peoples are not as well known as those of the valley of Mexico or the lowland Mayan area, and since Gage would be spending some time among them, it might be worthwhile to say something about their history. We are fortunate in possessing a good deal of aboriginal Quiché and Cakchiquel literature, which was written down in the native languages by literate Indians shortly after the conquest, and later translated into Spanish and other European languages. The best-known of these treasures is the *Popol Vuh*, a collection of mythical and historical legends of the Quichés.

The Quichés, and the peoples related to them, were at one time under the control of the Toltec empire; indeed, they resided at this time near the great city of Tula, in the Valley of Mexico. Tula was the capital of a militaristic empire which had arisen, centuries before the Aztecs, on the ruins of the theocratic empire of Teotihuacan. Up to this time, the peoples we are discussing seem to have been barbarians. Under the rule of the Toltecs, to whom they paid tribute, and under whom they served as mercenary soldiers, they became civilized.

Eventually they left Tula, setting out to conquer distant lands, and establish an empire of their own. First they attempted to conquer the Nonoualcas and Xulpiti of Southern Vera Cruz and Tabasco; but, while they had some success in this venture, they were eventually defeated by those who lived in the city of Zuyva. This 'formidable city' (to quote their chronicles) was located, according to some interpreters of these stories, on an island in the Laguna de Terminos in Campeche. Disorganized and depressed (*The Annals of the Cakchiquels* says 'we stripped off our feathers and we took off our ornaments'), they gave up their attempt to conquer the rich coastal area, and pushed up into what are now the Guatemalan highlands.

Here they employed Toltec methods of conquest: the use of human sacrifice to inspire terror (a device which the Aztecs used to the point of systematic atrocity), the performance of conjuring tricks which made the unsophisticated villagers think they possessed magic powers, and the more prosaic military arts of fortress-building and road-making. Once they had conquered the local peoples, they decided to legitimize their rule by sending emissaries to the two centres of Toltec rule at that time: the empire had split, and there was one centre in Yucatan, and one in the Valley of Mexico. They were, in effect, accepting the overlordship of the Toltecs, in exchange for Toltec recognition of their conquests. The emissary to Mexico apparently accomplished nothing. But the emissary to Yucatan returned with the insignia of Toltec royalty—the canopied throne, the royal flutes and drums, the feathered banners and head-dresses, and the jaguar claws.

Now, their sovereignty over the natives confirmed, the

conquerors fell to warring among themselves. By the time the Spaniards arrived, the Quichés and Cakchiquels were masters of the Guatemalan highlands, almost half of present-day Guatemala, and the Quichés were beginning to dominate the Cakchiquels.

Although their language belonged to the Mayan stock, the culture they established was Toltec in most of its essentials, with some admixture of elements from the Mayan lowlands, and from other Mexican cultures such as those of the Mixtecs and Zapotecs. The capital of the Quichés, Utatlán, had, the conquistadores tell us, twenty-four large stone houses in which the nobles dwelt. Each of these houses consisted of a large single room with a straw roof, raised on a platform about two yards from the ground, with a portico in front, and they were built around small court-yards, so that they faced each other. In these court-yards were held their religious dances. There was also, in the town centre, the palace of the ruling family, a pyramid temple, and a court for the ceremonial ball-game, which, like the tournaments of the European Middle Ages, was reserved for the nobility. To the conquistadores the architecture appeared pleasing, but undistinguished; and the sculpture was crude, compared to the beautiful stonework of the more highly civilized areas. It seems to have been a provincial variant of the great ceremonial centres of Mexico.

The town, which was built on a high mesa surrounded by ravines, had two entrances. One either climbed a steep staircase of twenty-five steps, or crossed a narrow causeway, laid with stones. Archaeological excavations have shown that these towns of the Guatemalan highlands had an extraordinary number of stone staircases, pyramids and buildings on high places, which made them very easy to defend against attackers.

In Gage's time there was nothing left of the old élite culture; the gods of America, like the old gods of Europe, had been conquered by Christ. No Christian can regret this: indeed, even if one were not a Christian, it would be naïve to regret it. The old civilization was tired; like other senile civilizations, it had become bloodthirsty, and was tearing the flesh of its own children between its rotting teeth. The Aztecs, like the Nazis, were a symptom of senile dementia. All over Mexico and Central America, civilization was falling to pieces. This is one reason why the Spaniards were welcomed. It is one reason, among others more exalted, why Christianity was welcomed.

But one of the unfortunate results of the conquest was that, while the terrorism and tyranny of the ruling class was destroyed, its high culture was destroyed also. Nothing was left but the ageless superstition of the Indian peasant; it was this, a force which could not be argued with, which could hardly be conquered even by the greatest and most discerning love, which the priests (and few of them were up to the task) had to cope with from day to day. It was in such an environment that Gage would be preaching the Word of God, and administering the sacraments. He would discover the tenacity of the old superstitions, and face supernatural terrors he was in no way prepared to face.

XIV—*In Santiago de Guatemala*

So renowned was Santiago de Guatemala that Gage had expected something as impressive as Mexico; but, in fact, he found himself in the city almost before he knew he was there. He did not enter through a gate, or pass over a bridge; no watch or guard stopped him with questions about where he was from, or where he was going; and he did not suspect he was in a city until he came upon a newly-built church standing near a refuse dump, with some mean houses near-by, some thatched and some tiled. He asked a passer-by what town he was in, and the answer came that he was in the City of Guatemala. The church, which was named San Sebastian, was the only parish church in the city. Otherwise, Santiago de Guatemala contained only its cathedral and the cloisters of the various orders. Farther on, Gage came to a broad street with houses on either side. Eventually, he arrived at the Dominican cloister, a proud and stately build-ing. He circled it on his mule, looking for the back entrance. When he found it, he alighted, knocked, and inquired for the Prior.

The Prior was Master Jacinto de Cabañas, who was also chief Master and Reader of Divinity in the College of St Thomas Aquinas, which was attached to the cloister. He had been brought up in the Spanish province of Asturias, which was visited by many English ships, and as a result he had become something of an Anglophile. For this reason, among others, he welcomed Gage, and promised to do all he could to assist him.

At the college, under Master Cabañas, Gage began to devote himself to serious study. He did so well that within three months he was nominated to defend, in a public debate before the entire college, a proposition on a subject which was, at that time, very controversial, and on which Master Cabañas had strong views.

The subject of debate was whether or not the Mother of the Lord had been born without the stain of original sin. The Jesuits, Franciscans, Scotists and the Spanish theologian Suárez maintained that she had been so born, that is, they upheld the doctrine of the Immaculate Conception. But the Thomists maintained, after Thomas Aquinas, that she, with all the race of man, had been born in original sin. Gage and Cabañas took the Thomist position.

The argument was in the medieval tradition of scholastic debate, which was still considered, in a neo-Hispanic culture soaked in the dark and heady wine of the Counter-Reformation, the only way to truth. A rank jungle of words and fine logical distinctions, a delicate green tracery of quibbles, sterile of fruit, but richly productive of leaf, grew up over the heads of the debaters, and in this verbal shrubbery major and minor premises sang like big and little birds.

The Jesuits stamped, clapped and shouted. They declared they were not surprised that Gage, an Englishman and born among heretics, should defend such a heresy; but they were surprised at hearing it defended by Master Cabañas, who had been born and educated in Spain, and was Chief Reader at the college.

Gage was not out to impress the Jesuits, though, but his fellow Dominicans. In this he succeeded. So impressed were they that when Michaelmas came round again, they made him a Reader of Arts at the College of St Thomas Aquinas. Gage thought enough of this honour to preserve the letters patent long after he became a Protestant minister. The document read as follows—

'Friar Juan Ximeno, Preacher General and Prior Provincial of the Province of Saint Vincent of Chiapa and Guatemala, Order of Preachers—whereas our Convent of Santo Domingo of Guatemala wanteth and standeth in need of a Reader of Arts: By these presents I do institute, name and appoint for Reader Friar Tomás de Santa Maria for the great satisfaction which I have of his sufficiency. And I command the Prior of the aforesaid our convent, that he put him in full possession and enjoyment of the said office. And for the greater merit of obedience, and under a formal precept, In the name of the Father, and of the Son and of the Holy Ghost, Amen. Dated in this our Convent of Chiapa la Real, the 9. of Feb. 1627. And I command these to be sealed with the great seal of our office.

Friar Juan	By the command of our
or	Reverend Father Friar
Ximeno Pal.	Juan de Santo Domingo
	Notary

I notified these letters patent, unto the contained in them the 12. day of the month of April, 1627.
Friar Juan Batista Por.'

Not everybody was as pleased by Gage's promotion as he was. The Creole party was much offended at seeing such an honour bestowed on a newcomer to the province. This opposition, though, and the support of his *gachupine* friends, only made him the more eager in the pursuit of learning, that he might fulfil the expectations which his best friends had of him. Like the other friars, he was expected to rise at six; but in the three years he was in that convent, he would often, sustained by a cup of hot chocolate, study until one or two in the morning. Indeed, he was reluctant to preach or to hear confessions, lest his studies be interrupted.

His superiors did not allow him to remain in seclusion very long. Impressed by his eloquence and learning, they demanded that he get out and preach. He was licensed to preach in the churches of the valley, that valley of Guatemala which was, as he discovered when he emerged blinking from his cell, most beautiful.

But first we should look at the beauties of the cloister in which he had lived for three years.

It was a rich convent, drawing its revenues (thirty

thousand ducats a year net) from a water-mill, a maize-farm, a ranch, a sugar plantation, a silver mine, and several Indian towns. Before the high altar in the church there was a picture of the Virgin worked in silver, a saint's statue with six lamps of silver burning before it, and another silver lamp which was so heavy that it took three men to pull it up with a rope. The treasure of the church and cloister, Gage estimated, was worth in sum a hundred thousand ducats.

A wide walk ran before the church, and this connected the college to the cloister. In the cloister there was a spacious garden: here there were two ponds, stocked with fish and visited by water-fowl, in which a fountain played continually. There were two other gardens, one of which contained a pond, surrounded by a wall and paved at the bottom, which was a quarter of a mile long. Here there was a boat for the friars' recreation; they would often fish from it, since this pond was also well stocked with fish.

Next to the friary was the convent of the Conception, and here there dwelt a young nun who was the wonder of the city, as much for her wit and accomplishments as for her beauty. Her name was Sor Juana de Maldonado y Paz, and she was the daughter of a judge. Sor Juana was particularly noted for her ability to invent verses *ex tempore*, a talent which greatly charmed the Bishop. She was clearly his favourite, and he was trying his best to make her the abbess, in spite of the opposition of the older sisters.

She was her father's only child, and the judge had spared no money in the adorning of her apartments. Thus she had rich and costly cabinets faced with gold and silver. The religious paintings in her rooms were set in black ebony frames with gold and silver corners, and her statues of saints glittered with jewels. Her private chapel, which was hung with tapestries, had an altar decked with jewels, candlesticks,

crowns and lamps, and covered by a canopy embroi-
dered with gold. In her closet or parlour she had a small
organ and a number of other musical instruments: here,
sometimes alone and sometimes in consort with her best
friends among the nuns, she would entertain the Bishop
with music. Eventually her father built her a new set of
apartments, containing rooms, galleries and a private garden walk, and he assigned six negro maids to wait upon
her.

As to the city, it was, as we have seen, not very large. The
city proper consisted of some five thousand families, and
attached to it was an Indian suburb of about two hundred
families. It was a fairly new town. The old capital, Ciudad
Vieja, which was about three miles away, had been destroyed by flood in 1542: one of the two volcanoes near the
city was split apart by an earthquake, and the water of a
lake in its crater came rushing down on the city, inundating
it, and destroying many of the houses. The capital was removed to the city in which Gage now was: Santiago de
Guatemala, or Antigua, as it is called today. The river,
which still flowed from the mountain, now fed the orchards
and gardens of the valley.

The other volcano was still active, and the city was often
shaken by earth tremors, especially in the summer. Once,
before Gage had come, it had spewed out burning embers
and shot out boulders the size of a house, killing all the
vegetation around its sides and foot. On another occasion,
it had burned so intensely for three days and nights that
Master Cabañas had been able to read a letter in its glare,
though it was dark, and the volcano was three miles from the
city. But Gage had learned to live with the volcano, and
fears of earthquake and eruption did not mar his pleasure
in the city and its surrounding countryside.

The valley was rich and fertile. Small towns and villages crept halfway up the mountains which enclosed the narrow valley: roses and lilies bloomed all year round, and the trees were heavy with sweet and delicate fruits. The cattle were so numerous that they were slaughtered for their hides, which were shipped to Spain. The beef, Gage tells us, was good, though inferior to that of England: it sold at thirteen pounds and a half for the equivalent of threepence, between one-sixth and one-seventh of the English price of the day. At one fair which Gage attended cattle were sold at nine shillings a head; and there were so many wild cattle (or *cimarrones*) abroad that hunting parties were organized to kill them, simply to prevent their damaging the crops.

There were a number of wealthy merchants in the city; but five were notably rich, being worth about five hundred thousand ducats each. Of these five merchants, one, Antonio Justiniano, was 'Genoese born'; and two were Spaniards. The other two were 'Portuguese', thus, in all probability, Jews: Jews in Spanish possessions commonly referred to themselves, and were referred to, as 'Portuguese'. The names of these latter two were Antonio Fernández and Bartolomé Nuñez, 'whereof the first in my time departed from Guatemala for some reasons which here I must conceal'.

This is one of the most mysterious statements in Gage's book. Why should he conceal Fernández' reasons for departure? He conceals nothing else about the people he met in the New World: indeed, he gleefully expands on their peccadilloes and oddities. He owed no special debt of gratitude to Fernández, or at least he mentions none. His silence can hardly be due to shame. Had Fernández' departure been connected in some way with an incident which showed Gage himself in a discreditable light, he would probably not have mentioned it. The phrase 'for some reasons which here I must conceal' can hardly fail to suggest political or religious intrigue, and that of a kind involving England, since otherwise there could hardly have been a motive to 'conceal' the 'reasons' in a book published in England. And he adds, even more mysteriously, 'The other four I left there, the three of them living at that end of the city called *Barrio de Santo Domingo*, or the Street of St Dominic, whose houses and presence makes that street excel all the rest of the city, and their wealth and trading were enough to denominate Guatemala a very rich city.' Why is he so careful to indicate where their houses are to be found? He does not give us specific addresses in his description of any other city. This is specially interesting in view of the fact that his entire description of Guatemala is clearly intended to be of use to an invading army: Gage was later to propose to Cromwell that mainland America be invaded through that province. There is an oddity about the whole business, which is discussed in an appendix to this book.

The description, which is as detailed as Baedeker, need not be recapitulated in detail; but among the careful descriptions are some good stories which, among other things, give an excellent idea of the state of the negroes, or, as Gage calls them, the Blackamoors.

A number of escaped slaves—called *cimarrones* after the Spanish term for wild cattle—roamed the woods and mountains. Like the Maroons of Jamaica, they were organized loosely into bands, and, like them, they waged occasional war against the settlements. Usually, though, they confined themselves to attacks on mule-trains; and their methods were rather gallant and Robin Hoodish: they took only the goods they needed, and never harmed those escorting the trains unless they retaliated with force. It was an intelligent policy: as a rule, the escorts of the mule-trains gave them what they wanted without fighting, and often the slaves in the caravans ran away and joined them.

Not all the negroes were slaves. Some were free men, and a few had prospered. There was one, at a place called El Agua Caliente, who was rich in cattle, sheep and goats, and particularly well-known for his excellent cheeses. It was commonly felt that he could not have become so wealthy simply by running an efficient farm: he was, after all, only a negro, and could thus have had no head for business. It was believed that he had found some hidden treasure. He was summoned by the Chancery Court of Guatemala, to reveal the source of his income; but his answer was disappointing: he merely said that when he had been young and a slave, he had been fortunate enough to have a kind master, who had allowed him to make money on his own; that he had, by the exercise of thrift and prudence, accumulated enough money to buy his freedom, and after that a small house with some livestock; and that from that moment God had prospered him.

Clearly this was an unusual case. Few slaves had such kindly masters, though few had masters so cruel as the man we are about to describe.

One of Gage's friends was a miser named Juan Palomeque He was worth six hundred thousand ducats, but he lived like a peasant on milk, curds, hard biscuit and jerked beef. Though he owned good houses in the capital, he dwelt among his slaves, in a poor thatched house little better than a hut. He was a cruel and vicious man. When he saw, in the streets of the city, an attractive female slave whose bearing indicated that she still retained a degree of personal pride, he would purchase her so that he might degrade her with enforced debauchery. In this he was no miser: in order to buy a female slave who had strongly aroused this obscene itch to kill the soul, he would frequently pay much more than she was worth in the market. He destroyed the pride of his male slaves in a different way, by sleeping with their wives. He took great pleasure in torturing his male slaves, and, here again, he concentrated on those who showed independence of spirit. One of these, whose slave-name was Macao, he would hang up by the arms, whipping him until his back was a mass of red flesh; then, as a salve, he would pour boiling grease over the wounds. He had branded Macao with irons on his face, hands, arms, back, belly, thighs and legs. Such tortures were forbidden, of course, by the Spanish laws governing the treatment of slaves.

As to Gage's role in the matter—and he does not seem to have felt that more was demanded of him—he did take it upon him to counsel Macao against hanging himself. The word 'friend', applied to Palomeque, is Gage's own.

So far as the Indians were concerned—'The condition of the Indians of this country of Guatemala is as sad and as much to be pitied as of any Indians in America, for that I may say it is with them, in some sort, as it was with Israel in Egypt. Though it is true there ought not to be any comparison made between the Israelites and the Indians, those being God's people, these not as yet—' Palomeque had himself killed two Indians, and had bought himself off at the trial, as if he had killed two dogs.

It is not news, of course, that the Indians and negroes suffered under the Spaniards. It is common to blame Spain and the Roman Church for such horrors. Yet it is clear that the real culprits were the Creole landlords, and the merchants who dealt in slaves and dominated the sugar industry, in which the worst abuses occurred. The business headquarters of the slave trade was, after all, in Amsterdam, not Seville or Cadiz. Spanish law relating to Negro slavery was cruel enough, in the sense that slavery itself is cruel; but it did not sanction atrocities. The laws relating to Indians were less cruel; indeed, they might be described as 'paternalistic'. Had not the King, in 'The New Laws for the Good Treatment and Preservation of the Indians', promulgated in 1542, declared of the Indians—'we wish them to be treated as our subjects of the Crown of Spain, for that they are'? And had not Pope Paul III said in a Bull of 1537 that they were inspired by the devil, who claimed 'that the Indians of the West and the South, and other people of whom we have recent knowledge, should be treated as dumb brutes created for our service, pretending that they are incapable of receiving the Catholic faith'?; and in this same Bull he refers to racist doctrine as 'never before heard of'.

Spain might be accused, with some justification, of a quixotic archaism. In her Indian laws, she was attempting to transplant the feudal system, already out of date in Europe, to America. The concept was not unfamiliar to the Indians: the Aztec lords had been something between imperial bureaucrats and feudal lords. But the Creole *encomenderos*, newly-created 'aristocrats', did not understand the obligations of the lord to his serf, obligations which, when they were honoured, made feudalism tolerable in the main, and in some cases, better than tolerable. They only saw that they had been given the right to exploit the labour of others. The spirit of rapacity and 'free enterprise', still largely expressed in an agricultural and mercantile rather than a manufacturing economy, was growing. The Spanish Crown resisted the trend: in the eighteenth century, for example, the Crown established a forty-eight-hour work week in the mines, and granted to the miners a profit based on a percentage of the ore mined. A new class, though, was springing up in the New World—a class of sugar-planters, mine-owners, owners of textile factories and workshops of all descriptions, and slave-traders: there was only one way to make money and rise in the world, and that was to participate, more or less enthusiastically, in a moral climate whose two extremes were petty venality and atrocity. If the court and church were to blame it was because, in spite of their attempts to be just, they could not resist the wealth pouring in from the New World. Complaining bitterly of the cruelty and dishonesty of their subjects and adherents, they averted their eyes, and held out their hands, palms uppermost, to receive the money made by the practices they professed to abhor. They had been compromised by the sordid demands of empire: had they attempted to enforce fully the laws they had promulgated, there would have been a Creole revolution, and Spain, deprived of the treasure of the New World, would have gone bankrupt.

Thus, though the enslavement of Indians was forbidden by law, they were often treated as slaves. Gage had ministered to some whose pay had been many blows, a few wounds, and little or no wages. These had 'sullenly and stubbornly' laid down on their beds, and starved themselves to death.

'The Spaniards that live about that country (especially the farmers of the Valley of Mixco, Pinola, Petapa, Amatitlán, and those of the Sacatepéquez) allege that all their trading and farming is for the good of the commonwealth, and therefore, whereas there are not Spaniards enough for so ample and large a country to do all their work, and all are not able to buy slaves and Blackamoors, they stand in need of the Indians' help to serve them for their pay and hire; whereupon it hath been considered that a partition of Indian labourers be made every Monday, or Sunday in the afternoon to the Spaniards, according to the farms they occupy, or according to their several employments, calling, and trading with mules, or any other way. So that for such and such a district there is named an officer, who is called *juez repartidor*, who according to a list made of every farm, house and person, is to give so many Indians by the week. And here is a door opened to the President of Guatemala, and to the judges, to provide well for their menial servants, whom they commonly appoint for this office, which is thus performed by them. They name the town and place of their meeting upon Sunday or Monday, to the which themselves and the Spaniards of that district do resort. The Indians of the several towns are to have in readiness as many labourers as the Court of Guatemala hath appointed to be weekly taken out of such a town. These are conducted by an Indian officer to the town of general meeting, and when they come thither with their tools, their spades, shovels, bills, or axes, with their provisions of victuals for a week (which are commonly some dry cakes of maize, puddings of *frijoles*, or French beans, and a little chile or biting long pepper, or a bit of cold meat for the first day or two) and with beds on their backs (which is only a coarse woollen mantle to wrap about them when they lie on the bare ground), then they are shut up in the town-house, some with blows, some with spurning, some with boxes on the ear, if presently they go not in.

'Now all being gathered together, and the house filled with them, the *juez repartidor*, or officer, calls by the order of the list such and such a Spaniard, and also calls out of the house so many Indians as by the Court are commanded to be given him (some are allowed three, some four, some ten, some fifteen, some twenty, according to their employments) and delivereth unto the Spaniard his Indians, and so to all the rest, till they be all served; who when they receive their Indians, take from them a tool, or their mantles, to secure them that they run not away; and for every Indian delivered unto them, they give unto the *juez repartidor*, or officer, half a real, which is threepence an Indian for his fees, which mounteth yearly to him to a great deal of money, for some officers make a partition or distribution of four hundred, some of two hundred, some of three hundred Indians every week, and carrieth home with him so many half hundred reals for one or half a day's work. If complaint be made by any Spaniard that such and such an Indian did run away from him, and served him not the week past, the Indian must be brought, and surely tied to a post by his hands in the market-place, and there be whipped upon his bare back. But if the poor Indian complain that the Spaniards cozened and cheated him of his shovel, axe, bill, mantle, or wages, no justice shall be executed against the cheating Spaniard, neither shall the Indian be righted, though it is true the order runs equally in favour of both Indian and Spaniard.'

The arrangement was, in fact, a *corvée*, or work tax. It might be compared to the *corvée* of seigneurial Quebec, and was not, in itself, any less just than our own money taxes. True, the average Indian would work on *corvée* three or four weeks in the year, as compared to three or four days for the average *censitaire*, or peasant, in Quebec. On the other hand, he was paid, while the *censitaire* was not. The Indian would be paid roughly half a crown for six days' work, a wage which appears ridiculous to us, but which, in terms of the purchasing power of money at that time, was not at all ridiculous, though it was small. For example, English agricultural labourers were receiving, at that time, weekly wages of from three to five shillings. Thus Indians were receiving half the high English rate and more than two-thirds of the low; and it must be remembered that the price of beef in Guatemala was one sixth to one-seventh of the English price.

Again we find that the laws of the Spanish Crown, which fixed the wages and attempted to regulate the working conditions of the Indians, were by no means unjust. This point cannot be over-stressed if we are to understand Gage's Mexico. The trouble did not lie in the law, but in the fact that the Creoles broke the law whenever possible. The

Spanish Crown had envisioned what might be called an enlightened feudalism, which was quite in the spirit of its archaising idealism: the Creoles turned it into a system approaching chattel slavery.

Some Spaniards, who did not have a full week's work for their Indians to do, would take advantage of their desire to return to the poor comforts of their homes; they would work them for half a week, and then say, 'What will you give me now, if I let you go home to do your own work?' Often the Indians would pay them for this privilege. Thus the land-owners would find themselves, after the Indians had gone, with their work done, and a small financial gain as well.

Some would whip the Indians for working too slow, or would wound them with their swords if they were rash enough to complain about their treatment. Some were even contemptible enough to confiscate their tools. Others would visit the wives of those who were assigned to them. Indian carriers—who sometimes carried loads of over a hundred-weight for a two- or three-day journey—would frequently be sent home, at the end of the journey, with no pay but blows, their masters having picked a quarrel with them to avoid payment.

There had been, and would be, insurrections, but they did not succeed. The imperialist principle of the *roi nègre* was applied with some skill. The descendants of the indigenous kings and nobles had their places among the minor gentry; the village chiefs were allowed to exploit their fellow-Indians in small ways; Indians with a sense of vocation or an intellectual bent were absorbed into the lower levels of the church hierarchy; and there was nothing to prevent an Indian's achieving a modest success as a farmer, rancher, or petty merchant. Thus potential leaders were diverted into working hard for small if real rewards. No Indian could carry arms, not even a bow and arrows; even a patent or nobility did not, as a rule, give him the right to carry a sword. Furthermore, the priests and friars calmed them whenever they were disposed to mutiny, persuading them to bear and suffer all for God's sake and the good of the Commonwealth.

The Indians had another consolation, and that was drink. '. . . if they can get any drink that will make them mad drunk, they will not give it over as long as a drop is left or a penny remains in their purse to purchase it.' They made their own *chicha*, and would go to insane lengths to strengthen it so that it would bring on the desired oblivion. In some towns they would put a live toad in the jar and close it up until the toad was thoroughly consumed. Once, in the town of Mixco, Gage and some officers of justice raided a house in which the Indians were drinking, and discovered four unopened jars of this liquor. When they were broken, and the contents were poured out in the street, the stink was such that Gage vomited on the spot.

They were, in fact, becoming 'drunken Indians'. The descendants of the war-like Tenochcas, the learned Texcocans, the subtle Mayans were being driven, melted and crushed down into that same state of clownish debauchery into which the tribes of North America would later fall. Out of a number of extraordinarily varied cultures and ethnic groups, the European was creating 'the Indian'. Certainly, drink had become to the natives—as narcotics have become to some of the negroes of the United States ghettos, as opium became to the Chinese coolie class in the decadence of the Chinese Empire, as marijuana, LSD and sexual drifting have become to many gentle and sensitive persons of all races in the United States—a way of escape from an intolerable situation. Certain Creoles were glad to help them escape. In spite of a law forbidding the sale of wine in Indian communities, there were a number of pedlars going from town to town, selling diluted wine strengthened with cheap and dangerous drugs. The local authorities turned their heads the other way, as they did in British Columbia in the 1860's, when the Indians were being poisoned or driven mad by a solution of red pepper in diluted alcohol, a drink from which a number of fortunes were made.

XV—*An Attempted Journey to Yucatan*

FOR three years Gage read in arts at the College of Guatemala. The more he studied, the more he became convinced that Roman Catholic theology was full of 'lies, errors, falsities and superstitions'. He was going through one of the typical crises of his time. He was becoming a Protestant. But in seventeenth-century Mexico, it was dangerous to be a Protestant: one might be burned at the stake.

The points which chiefly troubled him were seven: the transubstantiation of the consecrated elements during the sacrifice of the Mass, the existence of Purgatory, the merit o' man's works, free will, the priest's power to absolve from sin, the entire complex of beliefs and ritual practices associated with the saints and the Virgin Mary, and the infallibility o the Pope and his council—which last point, while it was not a dogma as yet, was often treated as if it was. In addition to these theological doubts, which might have assailed him anywhere, he was troubled by the greed, luxury and toleration of evil and oppression which he had found in the Mexican Church. Certainly, he may have been homesick as well. Yet we have no reason to question the sincerity of his doubts, or to deny that study, even of Roman Catholic authors alone, might well have strengthened them.

His doubts also strengthened his desire to return to England. He knew, he says, that there 'many things were held contrary to the Church of Rome, but what particulars they were I could not tell, not having been brought up in the Protestant church, and having been sent young over to St Omer'. Yet, while he wishes us to believe that he learned nothing of Protestant theology during this time, the evidence is that he was becoming acquainted with it.

He addressed petitions to his Provincial and to the President of Guatemala, asking that he might be allowed to return to Spain, there to be posted to the English mission. An order to this effect had, as we have seen, been sent from Rome; but it had not reached its destination. Gage's petition was rejected. There was an inflexible rule that all priests sent to the Indies must stay there ten years.

Gage therefore decided to request permission that he be allowed to leave the cloister, and become a priest in one of the Indian towns. It is clear enough that he intended, by delicately fleecing his parishioners, to make enough money to enable him to buy his passage home. Thus poor Gage, tempted by necessity and strong desire, celebrated his liberation from the chains of Roman Catholicism by rushing to embrace the same sin he had, up to then, so roundly condemned. At the same time, he wrote to a 'friend' in San Lucar in Spain, asking him to obtain a licence from the Court and from the General of the Order at Rome, which would assign him to England. The 'friend' was that same 'old crab-faced English friar', Pablo de Londres, who had tried to prevent his going to the Philippines in the first place.

Now at this time Friar Francisco Morán, Prior of Cobán in the Province of Vera Paz, was petitioning the President and Chancery of Guatemala for assistance in the discovery of a route from Guatemala into Yucatan. Many of the Indians there were still heathen, and had been attacking villages of Christian Indians. Morán was a friend of Gage; they had both studied, though at different times, in the cloister of San Pablo de Valladolid. Morán persuaded Gage to accompany him on this journey, in which many heathen would be converted, and, no doubt, much treasure discovered. 'I was not hard to be persuaded', says Gage with stupefying unctuousness, 'being above all desirous to convert to Christianity a people that had never heard of Christ; and so purposed to forsake that honour which I had in the University, in order to make Christ known unto that heathenish people.'

When they left, they took with them some money and gifts for the Indians, and an escort of fifty armed Spaniards. At the towns of San Pedro and San Juan, they added one hundred Indians to the party. For two days they travelled comfortably on mule-back, moving from one small Christian village to another; but as they approached the borders of Christian settlement, the roads turned to paths, and the paths narrowed and were blurred with grass, and finally melted into forest. Now their mules could not help them. They left them on the edge of the forest, and proceeded on foot.

For two days they moved up and down the slopes of wooded mountains, pushing through thickets and stumbling over rocks and fallen trees. At night they mounted a guard, but there were neither friends nor enemies to be seen. They found wild fruits in the mountains, and in the valleys and hollows, where the springs and brooks were, they found cacao and achiote trees. On the third day they came on a low valley with a shallow river running through it, and here they

123

found a few maize-fields. They drew together, and kept careful watch as they travelled.

Suddenly—and such discoveries can be very sudden in deep forest—they stumbled on a tiny hamlet of half a dozen thatched huts. Most of the inhabitants were away: they found only two men, three women, and five young children. When the Indians saw the Spaniards they began to weep aloud and cry for mercy. At this, Morán, who had a smattering of their language, tried to comfort them, and the Spaniards gave them some food and clothing. The Indians were forced to accompany the expedition, the hope being that they would lead the Spaniards to a treasure, or another village at least; but they refused to speak a word.

Later, they found some human tracks, and they followed them through the forest until they came, near dusk, to another hamlet, this time consisting of a dozen huts. It contained about twenty men, women and children. The expedition confiscated their bows and arrows, ate up most of their stock of plantains, fish and venison, and rested there the night. The next day they sent out some scouts, who returned to say that they had found other huts and garden plots, but no Indians, because they had all fled into the forest.

The next day they moved forward again, heading for a group of plantations which were, their scouts told them, in open country; here they could move more easily, and see more clearly any dangers—human or animal—which threatened them. They were moving along the river bank from one tiny village to another. It was hot and humid, and they were beginning to weaken with fevers and dysentery. Now their prisoners grew more communicative: there was some gold, they said, in the river, and farther on there was a great lake. There were many thousands of Indians living on its shores, and they were very warlike, and good bowmen.

Some of the more bellicose members of the expedition were encouraged by this move; many, however, began to murmur against Morán, who had brought them so far and placed them in such great danger.

At midnight, as the Spaniards were sleeping—most of them in hammocks, but some on the ground—the watch gave the alarm. Enemies were approaching, to the number, they estimated, of about a thousand. The Spaniards and their Indian allies began to beat their drums and fire off their muskets and fowling-pieces; and at this the attacking Indians, scattered through the trees, gave their war-cries, 'which uproar and sudden affrightment', says Gage, 'added sweat and fear to my fever'.

Morán had come to him to confess, which did not make him feel any better. But Morán comforted Gage as best he could: there was, he said, little they could do; the soldiers were around the camp in a ring formation, and they would be wiser to stay where they were, in which case they might well survive, than to attempt to flee, which would certainly result in death.

As it happened, the skirmish did not last longer than an hour. The attackers disappeared into the night-bound forest, leaving ten of their number behind as prisoners. In the morning, the Spaniards found thirteen of their attackers dead on the ground. Five of the defenders were wounded, but only one of them fatally: he died of his wounds the next day.

Later the prisoners were questioned. They answered in a spirit of noble defiance. If the Spaniards did not go away, they said, six or seven thousand would come against them. They added that they were well aware that the Spaniards controlled the country round about them; all they wanted was the little portion which they now had. Their only desire was to live there in peace. If the Spaniards wished to see their country, and pass through it as friends, they were welcome; but if they came to fight and enslave the people, as their neighbours had been enslaved, they would die fighting rather than yield.

There was much debate among the soldiers. Some, with Morán, felt they should take the Indians at their word, and go peaceably through their country until they came to Yucatan; others thought they should fight; others argued in favour of turning back. Nothing was agreed upon that day; but they could not move, in any case, because of the sick and wounded. That night the Indians gathered around the camp again, but disappeared when they saw the Spaniards were waiting for them. In the morning they decided to turn back, but not until Morán had asked the Indians if they would let him return later, and pass through their country to Yucatan. If they agreed, he said, he would come with no more than half a dozen Indians, and would place his life in their hands. They answered that this was quite acceptable to them.

The next year, Gage tells us, he did return, as he had promised, and passed through to Yucatan without incident.

But now they turned back, retracing their steps, and as they went Gage's fever began to leave him. With them they carried some of the Indian children they had seized in the villages. These were baptized in Cobán, and assigned to Dominican cloisters, where they would be brought up as Christians.

'I remained after this for a while in Cobán, and in the towns about, until such a time as the ships came to the gulf; whither I went with Morán to buy wines, oil, iron, cloth and such things as the cloister wanted for the present. At which time there being a frigate ready to depart to Trujillo (some occasions drawing Morán thither) I took ship with him. We stayed not much above a week in that port (which is a weak one, as the English and Hollanders taking of it can witness) but presently we thought of returning back to Guatemala by land through the country of Comayagua, commonly called Honduras. This is a woody and mountainous country, very bad and inconvenient for travellers, and besides very poor; there the commodities are hides, *cañafistula*, and *zarzaparrilla*, and such want of bread that about Trujillo they make use of what they call *cassave*, which is a dry root that being eaten dry doth choke, and therefore is soaked in broth, water, wine, or chocolate, that so it may go down. Within the country, and especially about the city of Comayagua (which is a bishop's seat, though a small place of some five hundred inhabitants at the most) there is more store of maize by reason of some Indians, who are gathered to towns, few and small. I found this country one of the poorest in all America. The chief place in it for health and good living is the valley which is called Gracias á Dios, there are some rich farms of cattle and wheat; but it lieth as near to the country of Guatemala as to Comayagua, and on this side the ways are better than on that, therefore, more of that wheat is transported to Guatemala and to the towns about it than to Comayagua or Trujillo. From Trujillo to Guatemala there are between four score and a hundred leagues, which we travelled by land, not wanting in a barren country, neither guides nor provision, for the poor Indians thought neither their personal attendance nor anything that they enjoyed too good for us.

'Thus we came again to Guatemala, and were by the friars joyfully entertained, and by the President highly rewarded, and by the city called true Apostles, because we had ventured our lives for the discovery of heathens, and opened a way for their conversion, and found out the chief place of their residence, and sent before us those children to the city who witnessed with being in the cloister our pains and endeavours.'

XVI—*Mixco and Pinola*

THE journey to the border of Yucatan had been an adventure; but there had been no treasure, and Gage had not raised one real towards his fare home. He therefore fell back on his original plan: he decided to take advantage of the opportunities in the Indian towns around the City of Guatemala.

He managed to obtain a licence which made him resident priest in the towns of Mixco and Pinola. In these towns, and in the neighbouring towns of Petapa and Amatitlán, the Indians spoke Pokoman, a dialect of the Mayan language.

Thus, two weeks before Midsummer Day in 1630, he went to the town of Petapa, six leagues from Santiago de Guatemala, and set to work studying Pokoman under Friar Pedro Molina, who was well versed in the language, and was now too old to carry out his duties without assistance. First, Molina schooled Gage in the rudiments of the language, then he put him to work for a fortnight declining nouns and conjugating verbs. Vocabulary followed, and oral work. After six weeks of this kind of study, Molina gave him some sermons to translate. By Michaelmas, in a little over a quarter of a year, Gage felt confident enough to preach to the Indians on his own, and Molina was sufficiently pleased with his progress to recommend that he be assigned his own parish.

The towns of Mixco and Pinola, which he was to serve, were governed from the cloister in Guatemala, as were all the towns in the valley. Gage was enjoined to submit a quarterly account to the cloister, which was entitled to all

Mexico City in the mid-seventeenth century—an English conception

A religious procession in Mexico City, 1663

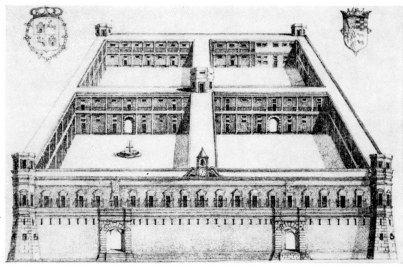

The viceroy's palace in Mexico City

moneys collected, after the resident priest had deducted his living expenses.

The sums involved were not inconsiderable. First, and perhaps least, each of the two towns had a common piece of land, sown with wheat or maize, whose profits were set aside for the priest, and turned over to him in monthly instalments. As for what else came in, let Gage tell of it: he kept very accurate accounts, and the figures are as illuminating to us as they were pleasing to him.

'Besides this monthly allowance, I had from the soda-lities of the souls in purgatory every week in each town two crowns for a Mass; every month two crowns in Pinola upon the first Sunday of the month from the sodality of the Rosary, and in Mixco likewise every month from three sodalities of the Rosary of the Virgin Mary, belonging to the Indians, the Spaniards and the Blackamoors, two crowns apiece. Further from two more sodalities be-longing to the Vera Cruz, or the Cross of Christ, every month two crowns apiece. And in Mixco from a sodality of the Spaniards belonging to San Nicolás de Tolentino, two crowns every month. And from a sodality of San Blas in Pinola every month two more crowns; and finally in Mixco from a sodality entitled of San Jacinto every month yet two more crowns, besides some offerings of either money, fowls or candles upon those days whereon these masses were sung; all which amounted to three score and nine crowns a month, which was surely settled and paid before the end of the month. Besides from what I have formerly said of the saints' statues which do belong unto the churches, and do there constantly bring both money, fowls, candles and other offerings upon their day unto the priest, the yearly revenues which I had in those two towns will appear not to have been small; for in Mixco there were in my time eighteen saints' images, and twenty in Pinola; which brought me upon their day four crowns apiece for Mass and sermon and procession, besides fowls, turkeys and cacao, and the offerings before the saints, which commonly might be worth at least three crowns upon every saint's day, which yearly amounted to at least two hundred three score and six crowns. Besides, the sodalities of the Rosary of the Virgin (which as I have said were four, three in Mixco and one in Pinola), upon five several feasts of the year (which are most observed by the Church of Rome) brought unto me four crowns, two for the day's Mass, and two for a Mass the day following, which they call the anniversary for the dead who had belonged unto those sodalities, which besides those days' offerings (which sometimes were more, sometimes less) and the Indians' presents of fowls and cacao, made up yearly four score crowns more. Besides this, the two sodalities of the Vera Cruz upon two feasts of the Cross, the one upon the fourteenth of September, the other upon the third of May brought four crowns apiece for the Mass of the day, and the anniversary Mass following, and upon every Friday in Lent two crowns, which in the whole year came to four and forty crowns; all of which above reckoned was as a sure rent in those two towns. But should I spend time to reckon up what besides did accidentally fall would be tedious.

'The Christmas offerings in both those two towns were worth to me when I lived there at least forty crowns. Thursday and Friday offerings before Easter Day were about a hundred crowns; All Souls' Day offerings commonly worth four score crowns; and Candlemas Day offerings commonly forty more. Besides what was offered upon the feast of each town by all the country which came in, which in Mexico one year was worth unto me in candles and money fourscore crowns, and in Pinola (as I reckoned it) fifty more. The communicants (every one giving a real) might make up in both towns at least a thousand reals; and the confessions in Lent at least a thousand more, besides other offerings of eggs, honey, cacao, fowls and fruits. Every christening brought two reals, every marriage two crowns; every one's death two crowns more at least, and some in my time died who would

leave ten or twelve crowns for five or six Masses to be sung
for their souls.

'Thus are those fools taught that by the priest's singing
their souls are delivered from weeping, and from the fire
and torments of Purgatory; and thus by singing all the
year do these friars charm from the poor Indians and their
sodalities and saints an infinite treasure wherewith they
enrich themselves and their cloisters, as may be gathered
from what I have noted by my own experience in those
two towns of Mixco and Pinola, which were far inferior
yet to Petapa and Amatitlán in the same valley, and not to
be compared in offerings and other church duties to many
other towns about that country, which yet yielded unto me
with the offerings cast into the chests which stood in the
churches for the souls of Purgatory, and with what the
Indians offered when they came to speak unto me (for they
never visit the priest with empty hands) and with what
other Mass stipends did casually come in, the sum of at
least two thousand crowns of Spanish money, which might
yearly amount to five hundred English pounds. I thought
this benefice might be a fitter place for me to live in than
in the cloister of Guatemala, wearying out my brains with
points of false grounded divinity to get only the applause
of the scholars of the University; and now and then some
small profit . . .'

Gage had made a careful study of the account books of
Mixco and Pinola. He had discovered that his immediate
predecessor had remitted, of the above-mentioned two
thousand crowns per annum, only four hundred to the cloister in Santiago de Guatemala. Those who had preceded him
had sent little more. Gage, with the most charming effron-
tery, proudly told his superiors he could do better: by the
practice of careful economies, he could promise to give the
cloister four hundred and fifty per annum. The Prior was
overwhelmed by such evidence of Gage's devotion. He told
him in return that he would never want for wine, but should

send his order in to the cloister every month. Furthermore, he could have a complete change of clothing once a year. Gage concludes:

'And here I desire, that England take notice how a friar that hath professed to be a mendicant, being beneficed in America, may live with four hundred pounds a year clear, and some with much more, with most of his clothing given him beside, and the most charge of his wine supplied, with the abundance of fowls which cost him nothing, and with such plenty of beef as yields him thirteen pounds for threepence.'

Gage now settled down to the comfortable life of a Mexican country priest. He had two houses, one in Mixco, and one in Pinola. He does not describe them explicitly, but scattered references in his narrative give us some idea of what they were like. His house in Mixco, which adjoined the church and a garden, had the thick walls so common in Spain and its colonies, and so suited to hot climates. It had a 'long gallery', probably more like a hall than the stately room the phrase suggests. His chamber was upstairs. The house at Pinola seems to have been of one storey, with four or five rooms, a small storage room, and a garden. There were servants living in each house.

He tells us nothing of the furnishings: Gage only mentions such things when he wishes to draw attention to the extravagance of others. We must therefore assume that it was furnished in the style typical of its time and place. Seventeenth-century Mexican and Central American houses were not, as a rule, cluttered with furniture. Even principal rooms might contain only walnut chairs and cushions, with a rug on the floor and hangings on the walls. Dining-rooms were usually furnished with a wooden table and benches. A bedroom might contain a canopied bed with turned and gilded columns, and those beautiful chests and coffers which were used instead of wardrobes and chests of drawers. Perhaps (we know this was the case in Gage's bedroom) there might be a writing table: wealthier persons would possess an elaborate escritoire. But though the pieces were few, they were richly worked. A massive virility of design, which in inferior pieces became a rustic heaviness, was combined with ornate surface decoration: in Mexico, not yet won over to the High Baroque style now fashionable in Spain, this carving was Renaissance, 'plateresque', in feeling. The furniture of the time had authority, if not delicacy. Each piece asserted its place in the room, and the effect, if the furniture was in the best tradition, was not one of bareness, but of a massive splendour.

Once Gage had established residence, he began to think again of ways in which he might return to England. He had become friendly with a certain Isidro de Zepeda, a Seville merchant and sea-captain who frequently passed through the valley. Gage had sent messages through him to his friends in Spain, though without much effect. Finally he threw himself on the captain's mercy, asking him to carry him in his ship to Spain. This the merchant refused to do, because of the danger it would place him in, if a complaint were made to the President of Guatemala. Rather, he told Gage, he should stay where he was, and accumulate money, so that he could return after his ten years with some wealth, and the approval of his superiors as well.

Gage was to spend five years in the towns of Mixco and Pinola. He would make a few friends, among them the 'Blackamoor' Miguel Dalva. It would be a lonely life, but not a poor one. In his time there were four great disasters, all of which proved to be of some profit to him.

In his first year there was a plague of locusts. When they came, it was in such a great cloud that they obscured the face of the sun; and they hung so thick on the branches of the fruit trees that they tore them away from their trunks. Gage, riding along the highway on his mule, had to cover his face as if he were in a snow storm. Where the locusts alighted, all green disappeared. The peasants were in a continual clamour for wafers with portraits of the saints impressed upon them, and these they buried in the fields, after Gage had blessed them, to drive the locusts away. Most of the images in the Mixco church were carried in procession to the fields. One cannot expect to have miracles for nothing; and, what with the masses, the processions and the wafers, Gage did well that year.

The next year there was a plague. Ninety died in Mixco, and in Pinola over a hundred. For every person over eight years old who had died, Gage received two crowns: the money would be used to finance a Mass, which would help his soul out of Purgatory. These deaths created an economic crisis: the Creole landowners of the valley, who used the Indians as farm labourers, began to worry about the reduction of the labour force. In the true spirit of enlightened self-interest, they ordered a numbering of the people, and forced into marriage all those of twelve years or over. Gage married over eighty couples that year. Each marriage brought in a little more money.

After the plague there were great storms. Many fields and cottages were flooded, and two houses and a chapel were destroyed. The high altar at Petapa was struck by lightning, and two travellers and a friar were killed by thunderbolts. Gage's own house was struck by lightning; but though two calves in his yard were killed, and he was knocked unconscious, he escaped serious injury. To protect the people from thunderbolts, there were more processions with images of the saints—and more money for Gage.

There was an earthquake the following summer. It was not serious. But there was a great and quite justified fear of earthquakes in Guatemala, and it seemed best to bring out the images of the saints once again, in order to prevent a more serious seismic disturbance. Thus there were more processions, and more money for Gage.

Now there is no doubt that Gage—in his hatred of the Papists—does not hesitate to paint himself in dark colours, so that the Church of Rome may stand condemned in him. But it is also clear that where money was concerned, he had no scruples, provided it was obtained within the law. Where money was not concerned, he could be compassionate and understanding, even courageous. For example, he hauled a Spaniard to court for beating an Indian almost to death, and this at some risk to his own life, for the Spaniard threatened to kill him, and, indeed, went so far as to enter his yard with a naked sword. Certainly the Spaniard was apprehended, fined and imprisoned; but such acts of courage should be taken into account, when one is drawing up the balance sheet of Gage's life in Mexico.

XVII—*Witches and 'Nahualistas'*

THE Christianity of the Indians was, Gage tells us, an external thing. 'Most of the Indians', he says, 'are but formally Christians, and only outwardly appear such, but secretly are given to witchcraft and idolatry.' What had survived, it should be remembered, was not the high religion of the great civilizations, which had had its own profundity. In pagan days, the peasants and poor craftsmen had known little of the mysteries of the Mayan or Toltec religions, in which astrology, magic, the physical sciences, high philosophy, and a degree of true religious feeling were combined in a tremendous and complicated synthesis. Nor were they encouraged to investigate such matters; it was enough that the priests and lords said that such and such was so. They performed whatever roles they were assigned in the great rituals held in the ceremonial centres, and returned home to their little family shrines, to the propitiation of the spirits of game animals and domesticated plants, and to the gross and sometimes rudely beautiful rituals which had long preceded the building of the great temples, and would outlast them.

There was one element of the old civilization which the Spaniards, or any Christian people, no matter how anthropologically enlightened, would have had to suppress most rigorously, and that was human sacrifice. This would not have been difficult in itself, since much enlightened Mayan and Toltec, and even Aztec, opinion had always regarded it as a dubious, even an abominable practice. Some of the customs and rites connected with sex (such as polygamy, the phallic rituals associated with the goddess Tlazolteotl, and the licensed promiscuity of the *telpochcalli*, or schools for male commoners) were also repellent. But here again, the enlightened among the Indians also found them repellent, or at least, 'common'. Indeed, the professed sexual morality of the Aztec upper classes was harshly Puritanical, though their private practice was often—and this is hardly unusual—something else again.

Obviously, it had been no mistake to attempt to Christianize Indian culture: many of the Indians welcomed Christianity, and some looked upon its coming as a fulfilment of ancient prophecies. Nor was it a mistake to prevent such abominations as human sacrifice by force, or to use force to protect those Indians who had become Christians. The mistake, rather, lay in the attempt to destroy Indian culture as a whole. For, as everybody knows, the Spaniards—except for a few missionaries whose understanding of native ways exposed them continually to charges of heresy—saw the devil in everything the Indians did. In this, the first conquerors were not unlike the Puritans of New England. They stamped out the most harmless native customs, and the high intellectual culture of the upper classes, with the same passion that they expended on the abolition of human sacrifice. As a result, the Indians themselves lost any sense of moral discrimination so far as their own culture was concerned: they either rejected it entirely, and accepted the role of half-men in a totally Spanish culture, or they clung indiscriminately to everything they could preserve, even human sacrifice, which became, almost, a symbol of cultural survival.

Thus the records of ecclesiastical and civil courts during the first hundred years after the conquest tell of many trials for this crime. These rustic rituals lacked the pomp which might have lessened their horror in pagan days: they were now mere murders hastily and awkwardly carried out in deserted country churches, or caves, or the bush. The victims were no longer acquiescent, and the priest's promises that the victim would go straight to heaven were now less readily believed. In a pathetic and disgusting attempt to capture the 'power' of the new religion, the clownish and ignorant pagan priests of the villages would, in some cases, crucify their victims. (In some cases the victim would be given, half-contemptuously, the ceremonial name of Jesus Christ.) The noble and exalted elements of the native religion, already debased by centuries of strife, had disappeared. In their place was a clandestine cult which had much in common with the then flourishing witch-cult of Europe, and which, indeed, absorbed elements of European 'underground' beliefs, brought over by the white settlers and conquerors.

137

'In Pinola', says Gage, 'there were some who were much given to witchcraft, and by the power of the Devil did act strange things.' There was one old woman, Marta de Carrillo, who was suspected by many of being a witch, but when she was brought before the justice she was acquitted since he could find no sure evidence against her. This acquittal emboldened her to practice her arts even more openly: two or three died in Pinola as a result of her machinations. As Gage describes it, they literally withered away, declaring on their deathbeds that Carrillo had killed them, and that they had seen her spirit near their beds, threatening them, and wearing an angry and frowning expression. It was known to be her spirit, since none of the others standing around the beds had seen her.

The Indians were so afraid of her that they did not dare to lay a charge. So Gage himself warned the lord of Pinola, Don Juan de Guzmán, that if her actions were not looked into, she would destroy the town. Guzmán, upon receiving this warning, obtained a commission from the Bishop and another officer of the Inquisition which empowered Gage to make a careful and secret inquiry into her life and actions.

He found that many of the Indians had complaints against her. Most of the town affirmed that she was a notorious witch. It was said that, before she had been brought before the justice, her familiar had been a duck. This duck had followed her wherever she went about the town; it even followed her to church, waiting for her outside the door until she came out. This duck, they said, must have been her devil and familiar spirit, since, when she had it with her, and they set dogs on her, the dogs would not chase her, but ran away. After she had been brought before the justice, though, the duck was not seen with her any more. She had warned her familiar off, it was said, to avoid suspicion.

Though Marta de Carrillo was a widow, and in appearance one of the poorest people of the town, she never seemed to be short of money. During Lent, all the town came to confession, each person bringing his gift. The usual gift was one real; but Carrillo brought four reals, not to mention eggs, fish, a little bottle of honey and a live turkey. Gage was not impressed: he reasoned that she had done this to improve the opinion he had of her.

In her confession she spoke of nothing but trifles, none of which could be called sinful actions. Gage questioned her very closely on the matter of witchcraft, and especially about those who had, in dying, seen her about their beds, threatening them. She replied, weeping, that she was wronged.

How was it, Gage asked her, that she, with no sons to help her, and no apparent means of livelihood, had so much money that she could give more at confession than the richest of the town? How had she come by the turkey and honey, since she kept neither fowls nor bees? God loved her, she said, and had given all those things: the rest, she said, she had bought with her money. When Gage asked who she had bought them from, she said simply that she had bought them in the town.

Gage's answer to this was to urge her to repent, and to reject the devil and all his works; but she only answered like 'a saintly and holy woman', and begged him to give her the Communion with the rest the next day. I dare not do this, Gage answered, and he quoted her the texts about not giving the children's bread unto dogs, and casting one's pearls before the swine; it would be a great scandal, he said, to give the Communion to one generally suspected of being a witch. The old woman began to weep, saying that she had received the Communion for many years, but Gage, unmoved, was resolute in his refusal. He dismissed her, and she left in tears.

At noon, his work in the church done, Gage sent his servant to gather up the offerings. In particular, he was to have Marta de Carrillo's fish dressed for dinner. But when the cook opened the fish in the kitchen, it was full of maggots, and stinking. When Gage was told of this, he opened her gift of honey and poured it into a dish. It was full of worms. Since there had been nearly a hundred eggs offered that day, he could not identify hers; but, out of the eggs he used during the rest of that day, some were rotten, and some had half-formed chicks within them. The next morning, the turkey was found dead. As for her four reals, when he counted the money which had been given him that day, and of which he had kept (who can doubt him?) a careful mental record, he found them missing.

That night the clerk of the church came to Gage, begging him to give Communion to Carrillo, since he feared that, if he did not, some harm would come to him. But Gage refused again.

After his servants had gone to bed, Gage sat up late with his books, preparing his Communion exhortation for the next day. It was between ten and eleven o'clock; he was in his chamber on the ground floor. He heard the hall door fly open; somebody came in, and walked about. Then another door opened, and the foot-steps continued into a little room where the saddles were. Gage thought it might be his friend, the middle-aged 'Blackamoor', Miguel Dalva. Dalva had given him much help in the town; they were on very easy terms, and Dalva often stayed at his house. He called him by name, two or three times, but no answer came. Suddenly another door—it led into the garden—flew open. Gage was now very frightened; his hair, he tells us, was standing on end. He tried to call to his servants, but his voice stuck in his throat. Then, remembering the witch, he put his trust in

God; a degree of self-possession returned to him, along with his voice; and he began to call to his servants, and to bang on the door with his stick, since he was afraid to open it.

The noise woke his servants, and when they came to his door, he summoned up enough courage to open it. Had they heard somebody in the hall, he asked them? Had they heard the doors being opened? They told him they had all been asleep; they had heard nothing. One boy, however, said he had been awake, and had heard just what Gage had heard. Gage picked up his candle, and went into the hall. All the doors were shut, and the servants swore that was just how they had left them. Then Gage perceived that the witch had tried to frighten him, but could do him no harm. He commanded two of his servants to remain in his chamber with him, and went to bed.

Early the next morning he sent for the clerk of the church, and told him what had happened during the night. The clerk smiled. It was the widow Carrillo, doubtless: she had often played such tricks with those in the town who had offended her. After the Communion service that morning, some of the leading Indians of the town came to Gage, and warned him that Carrillo had sworn to do him some mischief or other because he would not give her the sacraments.

Faced with such evidence, Gage could doubt no longer. He sent old Carrillo to the City of Guatemala, with a number of witnesses, and all the evidence which he had found against her (not much, it seems). The President and Bishop commanded her to be put into prison, and there she died within two months.

Even stranger was the case of Juan Gómez. Gómez was the chief of the principal tribe of Pinola and the countryside surrounding it. He was nearly eighty, and was a sort of village oracle, whose advice was sought on all occasions.

Gage had always thought of him as a devout Christian; he never missed morning and evening prayers, and had given the church a good deal of money.

Very suddenly he was taken sick, and the mayordomo of the sodality of the Virgin sent for Gage at Mixco, asking him to come and hear Gómez' confession: Gómez wanted much to see him, and to be comforted by him in his hour of death. Though it was pitch-dark and raining hard, Gage set out to ride the nine miles to his house.

When he arrived at Gómez' house, wet to the skin, he went to the old man's bedside. Gómez' face was heavily bandaged, but through the wrappings he thanked Gage for his charity and the pains he had taken; he confessed his sins with tears, and in his whole manner showed only a Christian character, and a willing desire to die and be with Christ. Gage comforted him and prepared him for death, asking him finally how he felt. The old man replied that he was sick of old age and weakness. With this, Gage went to his house, and laid down to rest. Before he went to sleep, more Indians came to see him: Gómez was about to die, and desired extreme unction. Gage went once again to his house to perform this last rite for the living, and first rite for the dead. He anointed him on the nose, lips, eyes, hands and feet. As he did so, he saw that Gómez' flesh was swollen and discoloured; but he did not think of it, at that time, as coming from anything other than the sickness of his body. It was now daybreak, and he returned home. After he had slept for a while, more Indians came to his door to buy candles to offer up for Gómez' soul: he had died, they said, and was to be buried that day. Groggily, Gage went to the church, and in the churchyard he saw a grave being dug.

Near the church he met two or three Spaniards, who had come into Mass that morning. He began talking about Juan

Gómez. He told them what comfort he himself had received from Gómez' Christian confidence, and that he was sure he would be saved for his godly life, and, finally, that the town would be lessened by his going, since he was the chief guide and leader of the people, full of wisdom and good counsel. The Spaniards smiled at one another. You are much deceived, they said, if you think Gómez was a saint and a holy man. Gage grew somewhat warm at this. You are enemies of the Indians, he said, and you judge them without charity, but I, who know them well from their confessions, can judge them better. It seems, one of them said, that Gómez' confession has told you little about him; apparently you are ignorant of what they say in the town about the real cause of his death. This seemed so strange to Gage that he asked them to explain themselves.

The report was, they said, that Juan Gómez was the chief of the wizards and witches of the town. He commonly took the shape of a puma, or mountain lion, and as such would walk about the mountains. There was one Sebastián López, another old Indian and head of another tribe, who was Gómez' deadly enemy. Two days before Gómez' death, they had met on the mountain—Gómez in the shape of a puma and López in the shape of a jaguar—and there they had fought. Gómez, who was older and weaker, had been much bruised and bitten, and his death had been caused by his wounds. López was now in prison on the charge, and the two tribes were contending: the people of Gómez were demanding compensation from those of López, and threatening to make the matter known to the Spaniards if they were not satisfied. But they were unwilling to do this, preferring to agree and smother it up among themselves, that they might not bring discredit on the whole town.

Gage did not know what to believe now: could Gómez

have deceived him, even on his death-bed? He confirmed that López was in prison by going there himself. He also found out that the town officers, and the chiefs of the tribes concerned, had been considering the matter in the town-house all that morning. Then he called one of the officers of the town, a friend of his, to his house, and questioned him closely. At first the man was reluctant to speak, fearing the whole town would turn against him if he told the truth; but when Gage comforted him with promises that this would not happen, he consented to tell what had occurred.

The story he told was substantially the same as that Gage had heard from the Spaniards. Some of Gómez' friends, he said, would not care if his use of black magic and his dark religious hypocrisy should be known to the world; others, he said, who were as wicked as the two old enemies themselves, wanted the matter kept secret, lest they and all the witches and wizards of the town should be discovered. Gage tells us that this struck him to the very heart. Was it possible that the best people among the Indians, who spent all they could get by their labour upon the church and the saints, should be in league with the devil?

When he had dismissed the Indian, he went to the church, to see if the people were coming to Mass; but he only found two men preparing Gómez' grave. He went back home, much troubled within himself as to whether he should give Gómez a Christian burial. On consideration, he decided he should, since he had only the evidence of the one Indian and the Spaniards, who spoke from hearsay.

At this point, though, a deputation, made up of nearly all the town officials, came to his door. Juan Gómez must not be buried that day, they said; they had decided to call officers of the crown to view his corpse, and examine into the mode of his death, lest questions should rise afterwards, and they

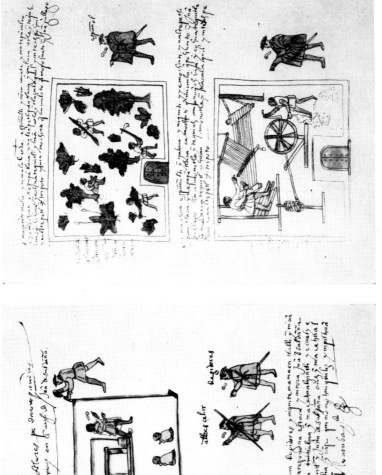

Spaniards taking over Indian lands

Indians at work on the land and in the *obrajes*

PANAMA.

Panama in the late seventeenth century

should have to exhume the body. Gage, pretending that he had heard nothing about Gómez, asked them why they wished to do this. They told him then that there were those in the town who had seen a puma and a jaguar fighting, and had presently lost sight of them, and that about the same time they had seen Gómez and López parting from each other. Gómez, covered with bruises, had immediately gone home to his bed, and had never risen again. He had declared on his death-bed that López had killed him, and for this reason they had put López in custody.

They begged Gage not to judge the whole town by these two wicked men, who had constant dealings with the devil, and that they were resolved to root out all such among them. Gage said that he did not think ill of such a laudable endeavour.

That night when the officer of the crown came, and he and Gage together examined Gómez' body, they found it bruised and scratched and bitten in many places. Thus, though a grave had been prepared for him in consecrated ground, he was not buried there, but in a common ditch.

Later López, who made little effort to deny the charges against him, was hanged in the city of Guatemala.

This was Gage's first encounter, though he was not fully aware of the deeper implications of the event, with the *nahualistas*. Like the Leopard-Men of Africa, they were members of a criminal secret society. They followed a perverted and ancient animal cult much older than the official religions of the Aztecs and Toltecs. They identified themselves with the jaguar, with whom, by means of revolting spells and rituals, they tried to establish a kinship. We should not be too quick to laugh at Gage and his villagers. Certainly the two old wizards could not, physically, have changed into jungle cats; but witchcraft is not, after all, concerned with the changing of physical objects. Travellers in Africa have reported seeing strange things which seem to indicate that it is possible for unscrupulous human beings possessing notable spiritual power to identify themselves in some way with animals, and, in a sense 'enter into' their wills.

It is the modern practice, among writers who preserve an open mind on questions of the supernatural, to recount what they have seen, without attempting to explain it. (Indeed, outside the writings of Swedenborg, there is no adequate theory of such phenomena.) If Gage does not confess the bafflement found in modern stories, this is because the theology of his day afforded an explanation of such phenomena as the result of a pact with Satan.

XVIII—*Breaking an Idol*

IN Mixco there were four brothers named Fuentes, who were among the wealthiest and most important of the town. They were very pious in their conversation; they were devoted to the saints, and scrupulously observed all the feast days; yet in private they were idolaters, continuing to worship their gods in the forests and caves around the town. It so happened that one day these, and others of their party, also rich men, were drinking *chicha* in the presence of some Christians. When they were drunk, they fell to boasting of their god, saying that he preached better to them than Gage did, and that he had told them to believe nothing that Gage preached to them of Christ, but to follow their forefathers, who knew how to worship the gods. The Christians were now very curious as to where that idol might be hidden. By pretending to yield to the Fuentes' persuasion to join them

146

in the 'old' worship, they managed to extract this information from them. When they were sober they thought better of it, and agreed to forget what they had heard, but could not forget it so far as to refrain from gossiping about it. Thus the news came to the ears of a Spaniard in the valley, who brought it to Gage.

Gage sent for the Indian from whom the Spaniard had heard the news. The Indian confessed that he had heard of such a thing, but that he was afraid to talk of it, fearing that if he did reveal where the idol was, the Indians would, by the power of their compact with the forces of hell, do him much harm. If you are a true Christian, Gage said, you would fight against the devil fearlessly, knowing that hell cannot harm you if you are with God, and have faith in Christ. Indeed, he added, the discovery of the idol will be a means of converting the idolaters, when they see what little power their false god has against the true God of the Christians.

Furthermore, Gage said, if the Indian did not tell him who the idolatrous Indians were, and where their idol was, he would have him sent to the City of Guatemala, where the justices would force the facts out of him.

At this, the Indian agreed to tell all he knew. The Fuentes, he said, had talked of a cave in a certain mountain, with a spring and a pine tree at its mouth. He could not be more specific: he had often been in that mountain, where he had seen two or three springs of water, but he had never been in any cave. Will you come with me, said Gage, to discover the cave? The Indian refused. He tried to persuade Gage not to go, lest, if the idolaters should be in their cave, they should kill them rather than allow themselves to be discovered. I will go with a guard, Gage said, and as to their false god, it has no power to hurt a Christian.

Gage decided to take with him the Indian, the Spaniard who had first told him of the idolaters, three or four additional Spaniards, and his mulatto friend, Miguel Dalva. He would not let the Indian return to his house, for fear he should reveal Gage's purpose in the town, thus forewarning the Fuentes to remove their idol. The Indian still refused to accompany him, and once more Gage threatened to send for the officers of justice. The Indian gave in, and Gage dismissed him in the custody of the Spanish informant, who was to keep close watch and guard over him until they met the next day.

That day he rode to Pinola and summoned Miguel Dalva, but without telling him why he wanted him back in Mixco. He also went to four Spaniards, his neighbours, and told them to be ready the next morning to go a little way with him 'for the service of God'. They should bring their fowling pieces, in case they should 'find some sport', and Gage would be responsible for the wine and meat. They readily agreed, thinking that, though he had talked of the service of God, he was simply inviting them to go on a hunting party. Gage was quite content that they should think that way. He went home, and had his servant prepare a good gammon of bacon, and some roasted and boiled fowls.

The next morning the party travelled towards the area in which the idolaters celebrated their clandestine rites. This was in a wood some six miles from Mixco, towards the town of San Juan Sacatepéquez. Shortly after they entered the wood, they came upon a deep *barranca*, or glen, with a stream running at its bottom. It seemed a likely place, but though they searched diligently, they could find nothing. (We must assume that by now every member of the party knew what they were hunting.)

They climbed up the side of the *barranca*, and, after careful

148

investigation, found a spring of water, but no cave. Evening
was now coming on, and some of the party began to speak
of returning home; but Gage thought it best to stay there the
night. If they returned home without success, he reasoned,
the idolaters might well suspect the purpose for which they
had been out. It would then be easy for them to rush to their
secret place and remove the idol. The company accepted his
reasoning, and they decided to stay in the *barranca* for the
night. When they had made a fire, they had a supper of cold
meat and hot chocolate.

In the morning they prayed for guidance, that they might
discover 'the cave of darkness and iniquity where lay hid
that instrument of Satan'. They entered the thick wood on
the side of a steep hill, and searched its south and north sides
without success.

Then, as they were descending the north side, they found
a poorly marked path, which they followed until they came
to another spring of water. When they had searched carefully about it, they found some pieces of broken earthenware,
and a piece of a chafing dish, such as the Indians used to
burn copal (Gage calls it 'frankincense') before the images of
the saints. It did not seem unlikely to the searchers that they
would use similar dishes to burn incense before their idols.
Further, they could identify the pottery as having come from
Mixco. Almost immediately, they found the pine tree which
the Indian had described to them. Near-by was the cave.
It was very dark within, but light at the mouth, and near
the mouth they found earthenware dishes, with ashes in them
—the remains of some copal incense. They struck fire, lit a
couple of candles, and went in.

At the entrance the cave was broad, and penetrated into
the mountain in a straight line; then it turned towards the
left. When they had followed this branch some eleven yards,

they found the idol. It was standing on a low stool, which was covered with a linen cloth. Though it was made of wood, it was black and shiny, as if it had been painted or smoked, and it was in the form of a man's head and shoulders. The idol had neither beard nor moustache; his expression was grim, his forehead was wrinkled, and his eyes were big and startled. However, 'we feared not his frowning look', says Gage. They lifted up the idol, and underneath they found a few reals, left there by the worshippers. On the ground about the image they found other simple offerings—single reals, a few plantains and other fruits, little jars of maize and honey, some half-burned candles, and little dishes on which incense had been burned.

These were precisely the things which were regularly given to the saints in Gage's church, some of which, he remarks dryly, were no handsomer than this idol. (Indeed, when Gage questioned the idolaters later, he expected them to say that it was the image of a saint, and thus to excuse themselves; but they did not; he was their god, they said, who had spoken and preached to them, and he was above all the saints in the country.)

The party now began to cut down the boughs of trees; they filled the cave with them, and stopped up its mouth. Then they came away, having wrapped the idol in cloth, and given it to the Indian to carry on his back. Gage thought it best to wait till nightfall before they entered Mixco, so that the idolaters might suspect nothing. He stayed at the house of one of the Spaniards until nightfall, and before he left asked him to warn all the Spaniards in that area to be at Mixco Church the next Sunday, because he had something to say to them and their negro slaves concerning their sodalities. He wanted the Spaniards there in case the idolaters— he did not know how many there were—should rise up against them; but he could not tell them why they were needed, for fear the idolaters, hearing of it, would stay away. Miguel Dalva, who was anxious to see the end of the business, stayed with him, and accompanied him home.

On the Sunday morning, before the service, he had Miguel Dalva carry the idol, still wrapped in its cloth, into the church, and hide it in the pulpit. Then he set Dalva to watch the church till the congregation came in, that none might see it, or take it away. He further commanded Dalva to be near the pulpit at sermon time, and to warn some negro friends of his, and those Spaniards who were aware of what was going on, to post themselves unobtrusively near the pulpit stairs.

That day the church was as full as Gage had wished it to be. The Spaniards were there, and the negroes, and among the Indians were the Fuentes and the other suspected idolaters, all quite ignorant of what Gage had in mind.

The Mass having been celebrated, Gage entered the pulpit for the sermon. He had chosen as his text the third verse of the twentieth chapter of *Exodus*—'Thou shalt have none other gods before me,' though in fact it was not the text of the day in the Roman calendar. The congregation looked at each other: they were not used to hearing sermons based on Old Testament texts. Gage then expanded on his text, his intention being to show that no finite creature could have the power of God the Creator, nor could any man do good or evil without His will or permission. If this was true of men, how much more was it true of inanimate creatures, such as idols.

The sermon half finished, he bent down in the pulpit, lifted up 'the black, grim and staring devil', and placed it on one side of the pulpit. He looked upon the Fuentes, and some of the others, and he saw that they changed colour, and looked at each other. Gage asked the congregation to see the god that some of them worshipped, and to ask themselves if they knew whence he had come, or what part of the earth he ruled. Some had boasted, he said, that that piece of wood had spoken and preached against Christ, and therefore they worshipped it as a god, and offered it honey, fruit and coins, and burned incense before it in a hidden place. Did they not show by this that they were ashamed to acknowledge him publicly, and did not his lurking in the darkness of the earth show plainly that he belonged to the Prince of Darkness?

Now Gage challenged the idol to speak for itself, or else by its silence to show it had no power. He taunted it, defying Satan, who had used it as an instrument, and daring the Devil to take it from the pulpit, if he could. Then he said to the Indians that if their god had power to deliver him from the execution he intended, which was to cut him in pieces and burn him, they should not believe the Gospel of Jesus Christ; but if the god had no power against him, the weakest instrument of the true and living God, then they should forsake their folly, and embrace Christ. He further promised them that, if they should renounce their superstitious worship, he would protect them from all punishments.

When he had concluded, he came down from the pulpit, and commanded the idol to be brought down after him. He sent out for an axe and two or three great pans of coals, and then he commanded that the idol be cut into very small pieces, and thrown on the burning coals; this was done, and the idol was burned before the congregation. At this the Spaniards cried out joyfully, *Victor, victor!* and *Gloria a nuestro Dios!*; but the idolaters did not speak a word. And with that the service ended.

Later Gage wrote the President of Guatemala, informing

him of what he had done. He also wrote the Bishop, asking him what should be done further, since he did not know how deep the rot of idolatry ran in the town. The reply from both was that he should discover as many of the idolaters as he could, and endeavour to convert them by fair and gentle means, to show pity on their blindness, and to promise them that on repentance they would be pardoned by the Inquisition, 'which, considering them to be but new plants, useth not such rigour with them which it useth with Spaniards if they fall into such horrible sins'.

Gage followed this advice. He asked the Fuentes to come to him privately. He told them how merciful the Inquisition was to them, expecting their conversion and amendment. But their demeanour was stubborn and angry, for, they said, not only they, but many others in that town and in the town of San Juan Sacatépequez, worshipped that same god. It is only a piece of wood, Gage said, so how can it be a god? We know it is only a piece of wood, one of them replied, and we know it cannot speak of itself; but we have all heard it speak. We believe God is in that wood, because if it speaks it must be something extraordinary. Certainly, it deserves more adoration than those saints in the church, who never speak to the people.

If you have heard it speak, Gage said, then it spoke of the Devil, not of God, because it preached against Christ, the only begotten Son of God. But they only answered that their forefathers had never known who Christ was until the Spaniards came into their country, but they had known there were gods, and they had worshipped them and sacrificed to them. For all they knew, they said, this god might have belonged to their forefathers.

Gage, seeing that he could not reason with them, told them to leave him.

From now on, Gage went in danger of his life. Twice the Fuentes, or some accomplices, entered his house and tried to get into his chamber. But the first time he awoke his servants, and they fled; and the second time, they were driven out by Miguel Dalva, who was now staying with Gage as a bodyguard. Yet they were always courteous when they encountered Gage in the street.

Then, almost three months after the burning of the idol, there came a messenger to say that Pablo, the oldest of the Fuentes, was dying, and that he wished Gage to convert him to the true religion. Gage was made very happy by this; he prayed that God would help him plead His cause, and he at once set out for Pablo's house.

When he had come to the door, and had entered one step, he found three of the Fuentes brothers, and some suspected idolaters, sitting about the room. Pablo was not there. Gage nervously withdrew his foot a little, and asked where Pablo was. They stood up and looked at him, but said no word. Gage, suspecting some treachery, turned about to leave, just in time to catch Pablo, the supposed invalid, creeping up on him with a cudgel in his hand, which he had lifted for the blow. Gage caught the stick with both hands as it came down, and while they were struggling with the stick, the Indians in the house ran into the yard and beset Gage, pulling him here and there and tearing his clothes. One ran his knife into Gage's hand to make him drop the stick (the scar remained with him all his life); but, because it was a public place, he did not dare to stab him in any vital part. When they saw he would not let go of the cudgel, another of the Fuentes took hold of it with Pablo, and they thrust it against his mouth with such force that they broke some of his teeth and his mouth filled with blood.

With this blow he was knocked down, but he soon re-

154

covered, and rose. They stood by, laughing at him, but did not dare to do him any more harm, for fear they should be seen. Indeed, they had already been seen. A mulatto slave girl was passing by, and heard Gage's cry for help—a hopeless cry, as it seemed, since all the houses in that neighbourhood belonged to the Fuentes. Seeing Gage covered in blood, she thought he had been mortally wounded, and ran off along the street, crying; 'Murder, murder in Pablo Fuentes his yard!' In the market-place and town-house she found the mayor and the officers of justice, who all came running, and with them two Spaniards with drawn swords.

Now Gage had to protect his assailant. The other attackers had run away, but he had held tight to Pablo. The Spaniards made to kill him on the spot; but Gage forbade this. Pablo Fuentes was taken off to prison.

That night the other idolaters and conspirators were taken, and also confined to prison. Later, they were brought to trial. They were whipped through the streets, and two of them were banished far away from their native town. All of them would have been banished, had they not pleaded with Gage to get them off, with promises to amend their lives. They confessed that the Devil had greatly tempted them, and that now they were prepared to forsake all their idolatries, and live thereafter as good Christians. Gage, much moved by their fears, interceded for them. Thereafter, he said, for as long as he lived in that town, he found a great change in their lives, which gave him every reason to believe their repentance was unfeigned.

Since we have evidence that Gage was capable of shameful and perfidious behaviour, let this good deed be recorded in his favour. He was genuinely concerned with the Indians' spiritual welfare; indeed, he loved them, and thought them 'of a good and flexible nature'. Part of his hatred of Roman

Catholicism, which caused him later to commit acts which seem unforgivable, must have received its final crystallization at Mixco. It was at Mixco he realized that the Mexican Church was Janus-faced. It had introduced Christianity; but it had also substituted the worship of the saints for that of the gods. In many cases, this had only involved the worshipping of idols with new names, depicted in a different art style, and representative of a different mystical cosmology.

XIX—*Gage Runs Away Again*

THE year was 1635. In this year, Gage received the letter he had been waiting for: he learned that the General of the Dominican Order in Rome had given him a licence to return to England. He was, indeed, weary of the New World: he had had little success in his efforts to help the Indians to advance beyond the merely formal Christianity they professed. Also, the Spaniard who was lord of the town of Mixco, a man little concerned with orthodoxy, was offended with him for having caused two of his town to be banished, and having caused an uproar over the affair of the idol. Accordingly, Gage wrote to his Provincial, who was then in Chiapa, asking for permission to return home. But the Provincial was now Jacinto de Cabañas, the very man who had sponsored, when he was Prior of the Guatemala cloister, Gage's debate with the Jesuits. Now he was more than ever convinced of Gage's value. Unwilling to lose so devoted a priest as Gage had shown himself to be (a priest, moreover, who spoke Pokoman) he appointed him vicar of the town of Amatitlán.

After a year, Gage tried again. By this time, Rev. Fr.

Jacinto de Cabañas had been replaced by Rev. Fr. Pedro de Montenegro, who simply transferred him from Amatitlán to the town of Petapa, where he stayed another year.

Finally, seeing that his superiors would never let him go, Gage decided to run away.

He could not afford to confide his plans to anybody: even his best friend might betray him. Yet it would have been a hard thing to run away alone, without some friend near him for the first two or three days' journey. There was another matter to be considered also: he had many things to sell, and he needed somebody to do this for him. He could think of no man better able to help him in this than Miguel Dalva, whom he knew, from long experience, to be honest and trustworthy. He sent for him, and, swearing him to secrecy, told him that he had to make a journey to Rome for his conscience' sake. He did not tell him the truth, that he was going to England, for fear that 'the good old Blackamoor' would, for the love he bore Gage, seek to prevent a journey from which his friend would never return. He would not be so alarmed by a trip to Rome: many, after all, had taken the same journey, and returned within two years.

Dalva wanted to go with Gage to Rome, but Gage refused, saying that the sea-journey would be too hard on a man of his age, and, furthermore, that as a free blackamoor in Europe, he might at any time be arrested on suspicion of being an escaped slave. Dalva could hardly fail to see the logic in this. Finally he asked only that he be allowed to accompany Gage to the seashore.

When Gage had sold his pictures, books and household furnishings, and had added to this the sum he had amassed during his twelve years in the New World, he found that he had nine thousand pieces of eight. There were no bank-notes or travellers' cheques in those days, and this was too heavy a load of wealth for a long journey. So he converted about four thousand crowns into pearls and precious stones, and the rest of the coins he laid up in bags or sewed into his quilt. Then he laid in some food for the journey.

He knew he would have to travel fast the first week. So he sent his chest on ahead of him, in the care of an Indian from Mixco whom he trusted well. Four days later he himself left, accompanied only by Miguel Dalva. He left when all the Indians were asleep, leaving nothing in his lodgings but old papers, and the key in the door. In this manner, Gage bade farewell to Guatemala.

He was fleeing towards Granada, on Lake Nicaragua. He had examined the other possibilities and rejected them. The ports on the Gulf of Honduras were the closest; but he knew that there he would, in all probability, meet people he knew. In any case, the ships left there so infrequently that orders to stop his escape might well arrive from Guatemala before he had embarked. To go through Mexico to Vera Cruz would be even more dangerous.

He left the week after Christmas, knowing that the ships commonly left Granada for Havana after the middle of January. He also had Miguel Dalva send a letter to a friend of his, which was in turn to be posted to the Provincial in Guatemala four days after his departure. In this letter he took his leave of the Provincial, asking him not to blame him or seek after him, since he had a licence from Rome to leave. And in order that the Provincial might not look for him in Nicaragua, he dated and subscribed the letter from the town of San Antonio Suchitepéquez, which was on the way to Mexico.

Until he reached the Rio Lempa, he was very careful not to travel through any large Spanish town, particularly one in which there might be a Dominican cloister. But when he

reached this river, 'some ten leagues from San Salvador', he
was able to relax, since

'This river is privileged in this manner, that if a man commit any heinous crime or murder on this side of Guatemala,
and San Salvador, or on the other side of St Miguel or
Nicaragua, if he can flee to get over this river, he is free as
long as he liveth on the other side, and no justice on that
side whither he is escaped can question or trouble him for
the murder committed. So likewise for debts he cannot be
arrested. Though I thanked God I neither fled for the one,
or for the other, yet it was my comfort that I was now
going over to a privileged country, where I hoped I would
be free and sure . . .

'We ferried safely over the river; and from thence went
in company with my Indian to a small town of Indians two
leagues off, where we made the best dinner that we had
done from the town of Petapa, and willingly gave rest to all
our mules till four of the clock in the afternoon.'

From the Rio Lempa he travelled to Realejo, and 'from
hence to Granada I observed nothing but the plainness and
pleasantness of the way, which with the fruits and fertility
of all things may well make Nicaragua the Paradise of
America'. Thus Gage arrived safely at Granada on the lake,
having caught up with his chest on the way. He was happy
enough when he got there, thinking that he had no more
land journeys to make until he should land at Dover in
England, and post up to London. It was time to think of
dismissing Dalva and the Indian; but Dalva refused to leave
him until he had seen him safely aboard ship. The Indian
was also willing to stay, but Gage would not allow this,
knowing he had a wife and children at home. The Indian
then declared himself perfectly willing to return home on
foot, offering to let Gage sell the mules to make what money
he could of them; but Gage, touched by this kindness, insis-
ted that he accept enough money to cover his expenses home,

and to keep himself for a while after he arrived. The Indian wept when he said goodbye, the third day after they arrived in Granada, saying that he would never see Gage again.

Gage and Dalva, left alone, discovered that the frigates would not be departing for a fortnight, so they decided to spend a day or two looking at 'that stately and pleasant tówn'. After that, they would take refuge in some nearby Indian town until the frigate sailed; for at that time many mule-trains were arriving from Guatemala, laden with indigo and cochineal for shipment to Spain, and there was always the possibility that somebody in the train might recognize them. Though Granada was on Lake Nicaragua, it was in effect a sea-port, since the ships passed from the lake to the sea by the San Juan River, or the Desaguadero ('channel'), as it was then called.

Granada they found to be a moneyed town, in which were some very wealthy merchants and a number of prosperous if less wealthy ones, most of whom traded along the Caribbean coast of Central and South America.

Its wealth was greatly increased while they were there. The Gulf of Honduras was infested, at that time, by the ships of Dutch pirates, and for this reason the merchants of Guatemala were sending their goods to Granada, that they might be shipped to Cartagena in what is now Colombia, and thence to Spain.

The town they chose to go to was a league from Granada; and here Gage was frequently entertained by the Mercedarian Friars, the most powerful order in the Indian towns of Nicaragua. Here also he made some inquiries about the journey to Cartagena. The answers he received were discouraging. Ships could cross the lake with ease, but the hundred mile journey down the Rio Desaguadero to the sea, was quite another matter. Sometimes the river voyage lasted

two months: because of the rapids, they had to unload the boats many times, and then load them again when the dangerous water had been passed. (The cargo was stored in a series of warehouses along the bank, while the boat, lightened of its load, passed through the rapids.) The gnats were intolerable, and passengers had been known to die of the heat before they reached the sea. It was distressing news indeed; yet he comforted himself that few frigates had been lost on the river, and that his life was, after all, in the hands of God.

Every so often Gage would go to Granada, to find out when the frigates were leaving, and to bargain for his passage. He found a captain, and settled with him for a place at his table. But, when they were within four or five days of departure, frustrating news arrived from Guatemala: it was ordered that the frigates should not go out that year. Apparently some English or Dutch ships were lying in wait off the mouth of the Rio Desaguadero. News like this was frightening not only to the merchants, understandably fearful of losing a fortune to the pirates, but to the President himself, who was responsible for the King's revenues, and would be held to account if, rashly ignoring due warning, he allowed them to depart on a ship which stood a good chance of being captured.

This left Gage at loose ends. He thought of the ship which ran between Realejo and Panama, but discovered that it had already set sail, as had the small ships from the ports of Honduras. Other passengers for Spain were in the same position; and when Gage consulted with them, they thought of hiring a frigate to carry them to Cartagena, but could not find a master willing to risk his ship at such a dangerous time.

Thus Gage haunted the quays of Granada, his hopes fading. At last he met a merchant, who gave him some advice. Gage should go, he said, to the port of Cartago in Costa Rica. There he would be sure to find a small coastal vessel bound for Portobello, laden with meal, bacons, fowls, and other provisions for the galleons. It was a long chance, and the journey to Cartago would not be at all pleasant: he would have to travel almost a hundred and fifty leagues, over mountains and through deserts, and he might, in the end, not encounter a ship. But anything seemed preferable to returning to Guatemala; even the long doubtful journey, and the waiting in Portobello till June or July, the months in which the galleons left for Spain. To three Spaniards he had encountered, the risk seemed worthwhile, and the four agreed to set out on the journey together.

Dalva would gladly have gone with him: indeed, he would have gone around the world with Gage, but this the latter would not permit. He thanked Dalva for what he had done, gave him money for the trip home, and dismissed him. The four companions set out from Granada on muleback, following the shore of the lake of Nicaragua.

On the second day out of the town, Gage had an adventure, which he records thus—

'The second day after we set out we were much affrighted with a huge and monstrous cayman or crocodile, which, having come out of the lake (which we passed by), and lying across a puddle of water bathing himself, and waiting for some prey, as we perceived after, when we, not knowing well at the first, but thinking that it had been some tree that was felled or fallen, passed close by it; when on a sudden we knew the scales of the cayman, and saw the monster stir and move, and set himself against us; wherewith we made haste from him, but he thinking to have made some of us his greedy prey, ran after us, which when we perceived, and that he was like to overtake us, we were much troubled, until one of the Spaniards (who knew

better the nature and quality of that beast than the rest)
called upon us to turn to one side out of the way, and to
ride on straight for a while, and then to turn to another
side, and so to circumflex our way, which advice without
doubt saved mine or some of the others' lives, for thus we
wearied that mighty monster and escaped from him, who
had we rid our straight way, had certainly overtaken us
and killed some mule or man, for his straightforward
flight was as swift as our mules could run—'

Was Gage's monster an alligator, a cayman, or that much
rarer beast, the *Crocodilus americanus*? Crocodiles and alli-
gators can move with some speed on land, and it is certainly
true that, since they can only run in a straight line, they may
be evaded by a zigzag flight. Perhaps Gage was so frightened,
never having encountered such an animal before, that he
thought it was moving as fast as his mule.

On the second or third day they left Lake Nicaragua, and
began to strike across rough and mountainous country
towards the city of Cartago. They had been informed that a
frigate would soon be setting out from the mouth of the River
Sueré (now called the Pacuaré). After four days' rest in
Cartago, they set out again. The country they were travelling
through was mountainous, with occasional valleys in which
Spaniards and Indians lived on isolated farms, growing a
little maize and breeding a few hogs. They arrived at the
mouth of the Sueré three days before their ship was to
depart.

The master of the frigate was very glad to have them
aboard. Indeed, he offered to carry Gage for nothing to
Portobello, declaring that he would be content if Gage
merely spent his time praying for him and the safety of his
ship. Not that the cargo of the ship would be very attractive
to pirates: it consisted only of honey, hides, bacon, meat and
fowls.

They weighed anchor, and headed towards the open sea. The mouth of the Sueré was rather dangerous: the river ran with a strong current, and the bay was shallow in some places, and crowded with rocks in others, but they reached open water without mishap.

When they had sailed no more than twenty leagues in open sea, they saw two ships making towards them. The passengers were very alarmed, especially since they could see that the captain, who suspected the ships to be English or Dutch, was looking rather worried himself. Their ship was not armed: the only weapons on board were four or five muskets, and half a dozen swords. It was a small, swift craft, which was some comfort; but they had not sailed more than five leagues further when they perceived that the ships were gaining on them, and that they were Hollanders.

Soon one of them, a man-of-war, fetched up, and fired a shot across their bow. They could do nothing but yield, and hope for mercy. 'But O, what sad thoughts did here run to and fro my dejected heart, which was struck down lower than our sail. How did I sometimes look upon death's frightening visage!' Even if Gage escaped death, what would happen to his 'treasure of pearls, precious stones, and pieces of eight, and golden pistoles'?

Gage had little time for Baroque meditations on death and fortune, since the Dutch, bristling with swords, muskets and pistols, boarded them almost immediately. But Gage and the other passengers felt a little better when they learned who the pirate captain was. He was Diego el Mulato, who had some four years earlier sacked Campeche. This was hardly a recommendation; but they knew that Diego had been born and brought up among the Spaniards of Havana. Thus he was preferable to a Dutchman, who could not be expected to show any mercy to Spaniards.

Diego was, as his name indicates, a mulatto. He was a proud man. When some Spaniards in Havana had whipped him for an offence he had not committed, he had stolen a small boat and taken it out to sea. There he had found some Dutch ships lying in wait for a prize, and he had offered himself to them, promising to serve them against his own countrymen.

So well had he done this that the Hollanders had grown to respect him. They had given him a Dutch girl as his wife, and had made him captain of a ship under Pie de Palo, or Wooden Leg, a Dutch captain much feared by the Spaniards.

There was little to seize on the ship—little, that is, except Gage's money, that money which he had amassed over such a long time, and at such peril to his soul. Diego took from him four thousand pieces of eight in pearls and precious stones, and almost three thousand more in coins. From the other Spaniards, he took some hundreds apiece. It was a richer haul than he had expected: as a result he took little interest at first in the cargo of the ship. Nor was he interested in Gage's other belongings. He did leave him his quilt, some books, some holy pictures, and his clothes, 'which I begged of that noble captain the mulatto'.

It is admittedly unusual for the victim to call the robber 'noble'; but in the continual, undeclared war which England was carrying on with Spain, pirates and free-booters had a strange and particular status. At the time he wrote his book, Gage was more concerned with pandering to the fixed ideas of the English public than with defending his own name; he is also perfectly willing to present this mishap as a Providential chastisement of his own greed, the only sin he admits in himself. Diego, who seems to have possessed the shoddy gallantry of the true seventeenth-century pirate, allowed Gage to keep these articles, which had little resale value, as if this were an act of extraordinary generosity. He added that he could hardly do otherwise than take his money and pearls, 'using that common proverb at sea—*Hoy por mi, mañana por ti*—today fortune hath been for me, tomorrow it may be for thee'.

Luckily, in the quilt which Diego el Mulato had so generously allowed him to keep, he had sewed up a few pistoles, and a few more were sewed up in his doublet: the sum came to almost a thousand crowns.

After the captain and soldiers had thoroughly examined the ship, and taken everything of monetary value, they began to think of the provisions it carried. With its cargo, the 'good captain' gave a dinner in his own frigate, to which Gage was invited. Many healths were traded back and forth. Among them, Diego drank one to his mother, asking Gage to see her and remember him to her, and adding that it was for her sake that he had, while taking from them all their goods, done it with such courtesy and good temper. And he added that, for Gage's sake, he would let them keep their ship, so that they might return to land and find some safer way of reaching Portobello.

After dinner, Gage talked alone to Diego. He told him that he was no Spaniard, but an Englishman born; he showed him the licence he had from Rome to go to England, and said he hoped that, since the English were not at that moment at enmity with the Hollanders, the good captain might give back what goods belonged to him. But his pleas made little impression on Diego, who had already taken possession of all the valuable goods in the ship; he said that Gage must suffer along with those among whom he had been found, and that he might as well claim all the goods in the ship as his own. Gage then asked him if he would carry him to Holland, that he might get to England from that country;

but Diego refused this also, saying that he went about from one place to another, and did not know when he would return to Holland. Furthermore, he said, they were ready every day to fight any Spanish ship they encountered; and if Gage should be on board when such an encounter took place, his men might, in hot blood, do him some mischief, thinking he would harm them by his testimony if the Spaniards should take them captive. Gage saw that there was no hope of getting back what he had lost, so he commended himself to God, and decided to be patient.

The rest of that day and part of the next, the Dutch pirates began to unload the cargo of the coastal boat into their man-of-war. They had changed their minds about seizing only 'legitimate' prizes, money and goods with an immediate cash value: they had decided that the fowls and bacon could be eaten as well as their own, that the meal would be useful to make bread, the honey to sweeten their food, and the hides to make them shoes and boots. In the end, they stripped the ship bare, leaving only enough food to satisfy the stomachs of the crew and passengers till they should make land, which was not far off.

'Thus they took their leave of us, thanking us for their good entertainment,' no doubt with all the contemptible flourishes of the pirate, who, like his contemporary the highwayman, or the modern Mafia gangster, adhered to a grotesque perversion of the honour code of the military aristocrat.

When the four travellers had returned to Sueré, the Spaniards of the towns in the area pitied their condition, and took up a collection for them.

The four, aided by this act of charity, made up their minds to return once again to Cartago, and thence to set out for a more practical port. On the way, they compared notes as to what they had saved. The three Spaniards who had accompanied Gage were in a somewhat better state than he was, for though they had lost all their money, and most of their best clothes, they had reserved some bills of exchange for money to be taken up at Portobello. Gage regretted that he did not have his money in bills like these, which had some of the advantages of the modern cheque. He had learned to be close-mouthed, and he did not tell them of the coins he had saved.

When they came to Cartago, they aroused much pity, and another collection was taken up. Gage officiated at Mass, and preached sermons wherever he went, and in this way he began again to accumulate a little money. The travellers agreed among themselves that they would look as poor and miserable as possible, so that the Indians and Spaniards on the way might be more disposed to give them money.

Nevertheless, Gage knew that in such a comparatively poor country, a country in which he was, moreover, unknown, he could not possibly accumulate enough to allow him to land in England with a full purse. For a moment he thought of returning to Guatemala to begin once again the laborious process of raking in money; but it seemed better to him that he should go home, even if he had to beg his bread on the way. He was by now, he tells us, thoroughly disillusioned with Roman Catholicism. Yet he continued to preach, fearful that the Spaniards might ask themselves why he refused to exercise his orders and function.

At Cartago, he inquired how he might get to Portobello. Once again, the news was discouraging. There were trains of mules which went over the mountains of Veragua to be sold in Panama. This was the only land traffic into that mountainous isthmus, and it was extremely dangerous, not only because of the crags, rocks and mountains which had to be crossed, but more specifically because the unconquered

Indians of the region regularly raided these mule-trains—not for plunder, since the mules themselves were travelling unladen, but out of pure hatred for the Spaniards. Only recently, he had heard, there had been such an attack, in which some Spaniards had been killed.

There was only one alternative, and that was to retrace their steps, to go to Nicoya in Nicaragua, and thence to the Gulf of Salinas, on the Pacific Coast of Nicaragua. Here they would be able to take ship to Panama. There was, after all, less danger that pirates would be waiting outside Panama, which was a Pacific port.

The way from Cartago to Nicoya was mountainous, and the towns and ranches were poor and insignificant. Nicoya itself, the administrative centre of that region, was, however, a pretty little town.

Gage tells a story about Nicoya which once again reveals that conflict between church and colonists on the question of the just treatment of the Indians, a conflict which seems to have been universal in Spanish America. It also shows that, in spite of its venality and money-grubbing, the church did, on occasion, act in accord with the doctrine it professed.

The *Alcalde Mayor* (this title was given to the governor of a city and its surrounding districts) was one Justo de Salazar. He entertained the travellers well, and provided lodgings for them. He also reassured them by saying that, while there were no ships in the Gulf of Salinas at the moment, he was sure that one would come from Panama very soon. It was that time of year when the coastal ships came for salt and other goods.

They had arrived in Lent. This was a good time, it seemed to Gage, in which to make money, because of the increase in the numbers of those who came to town for confession and Communion. He went to see the Franciscan who had 'the pastorship and charge' of the town, and an arrangement was made.

Now this friar had special reasons for desiring the assistance of Gage.

Some three weeks earlier, he had had a quarrel with de Salazar which had almost resulted in his death. The *Alcalde Mayor* was a notable oppressor of the Indians: he employed them in his and his wife's service as if they had been slaves, refusing to pay them their wages, keeping them from their homes and their wives, and commanding them to work for him on Sundays as on any other day. Accordingly, the friar commanded his Indians from the pulpit not to obey such commands, since they were unlawful. But Justo de Salazar was a hot-blooded man, skilled in fighting, and trained in the wars: he could not accept reproof from a mere friar on such matters as his treatment of Indians and his style of money-making. At first the quarrel was only verbal, and many hot words passed between them. But one day, words having ailed him, the *Alcalde Mayor* came to the friar's house with his sword drawn. The friar, who was as courageous as the *Alcalde*, stood against him, thinking that Salazar would not strike him, for fear of excommunication. Salazar was so provoked by this that he did what he had, perhaps, not expected to do: he lifted up his sword and struck at the friar. His first blow struck off two of the friar's fingers. There was no second because the Indians who served the priest pulled him by main force into his chamber, and shut the door on him, in order to protect him from his own courage.

Justo was indeed excommunicated; but since he was a man of wealth and influence, he managed to obtain a release from the excommunication by an appeal to the Bishop of Costa Rica. Not content with this, he malignantly complained to the Council of Guatemala against the friar; and there, using his friends and money, he caused the friar to be brought before the court. The upshot of it was that the friar was condemned to be removed from Nicoya.

The friar had not left when Gage arrived, but kept to his house and chamber, refusing to go to the church to say Mass, preach or hear confessions. He did have one assistant, but one was not enough at that time, since the church was thronged with 'many hundred Indians, Spaniards, Blackamoors and mulattoes', who had come up for Lent. Thus Gage became his second assistant, and for this he was rewarded with board and lodging at the friar's house, a crown a day for saying Mass plus the voluntary offerings of the people, and a fee for each sermon he preached. From the second week of Lent to Easter week, Gage made a hundred and fifty crowns.

The week before Easter, news came that a frigate was on its way from Panama to the Salinas Gulf. Shortly thereafter the master of the frigate came to Nicoya, which was, as we have seen, the chief town of the area. Gage and his three companions hastened to meet him, and made arrangements for their passage to Panama.

They travelled with him to the Salinas Gulf, and boarded the ship, which was laden with the usual commodities of the country—fowls, salt, honey, maize and some wheat. More important than these workaday goods, though, was the chief and richest cargo of the ship, a certain purple dye made by the Indians from the *Purpura* shellfish. This dye had been made, in pre-Columbian times, from the Pacific Coast of Mexico down to Peru: the process, which is complicated, resembles that involved in the manufacture of 'Tyrian purple'.

Once he was on the boat, Gage thought his troubles were over: he would be in Panama in five or six days, and his return home was only a matter of time. But on their first day out of the gulf, a storm arose, which drove them towards the south; after that, another storm drove them back towards the mountains of Veragua, north of Panama. Some said there must surely be an excommunicated person aboard, and that, if they could discover him, they would throw him overboard. They were becalmed, until 'a strong current' took them out to sea again. The water ran out, and they had nothing to drink for four days: Gage was reduced to drinking his own urine, and sucking on bullets. Nevertheless, the captain refused to stop at an island in order to pick up water, assuring them he would have them all safe and sound in Panama in a day. At this, some of the mariners, along with the three Spaniards accompanying Gage, drew their swords, and forced him to put into an island.

On the evening of the next day, they arrived at Puerto de Perico, outside Panama.

XX—*Portobello*

FOR fifteen days, Gage stayed in the Dominican cloister at Panama. He was, as always, very curious about the city; he had the obsessive and indefatigable curiosity of the good travel-writer. He noted that the city was governed, like Guatemala, by a president, six judges and a Court of Chancery, and that it was a bishop's see. It was more strongly fortified towards the Pacific than any other American port on that ocean, and the defences were armed with ordnance. Yet the houses themselves were as flimsy as those of Vera Cruz; lime and stone were hard to come by, and most of them were made of wood. The President's house, even the richest churches, were mere board constructions, covered with tiles. Because of the heat, the common clothing of the inhabitants was 'a linen cut doublet, with some slight stuff or taffeta breeches'. Gage adds, 'The Spaniards are in this city much given to sin, looseness and venery especially, . . . [and they] . . . make the Blackamoors (who are many, rich and gallant) the chief objects of their lust.'

Panama was thought at this time to be one of the richest places in America, but only in the sense that Vera Cruz was rich, that is, that wealth was in continual passage through it. Goods from the Atlantic trade came overland or via the River Chagres (which would, much later, become part of the Panama Canal system). Over the Pacific, Panama traded with Peru, the East Indies, and the western ports of Mexico and Central America. For all that, it was a small town, with some five thousand inhabitants, and eight cloisters of nuns and friars.

Gage, fearful of disease in a town noted for its high mortality rate, left Panama as soon as he could. One could travel to Portobello either by land or by water. But the mountains made travel difficult, and for this reason he decided to go by the River Chagres. He set out at midnight for Venta de Cruces, some thirty to thirty-five miles from Panama.

Having travelled by night to avoid the heat, he and his companions arrived, by ten the next morning, at Venta de Cruces, 'where live none but mulattoes and blackamoors, who belong unto the flat boats that carry the merchandise to Portobello'. Gage was well received by these people, who asked him to preach to them the following Sunday, and gave him twenty crowns for a sermon and procession. After he had stayed there five days, the boats set out again.

They had a difficult passage down the river. In some places, the river was very shallow, and their boats grounded on the bottom, whence they had to be released, with much poling and lifting, by the boatmen. In other places, the stream ran very swiftly through a tangle of over-leaning boughs, which they had to chop down. Luckily, a heavy rain swelled the river, and made their passage easier than it might have been.

The difficulty of navigating the Chagres was one of the strongest points in the defence system of Panama: the Spaniards relied upon it completely. This Gage and his companions discovered when, after twelve days, they reached the castle at its mouth. It was ready to fall to the ground. Apparently the Spaniards, thinking of the river as their ally, did not concern themselves with keeping the castle in good repair. The Governor of the castle, who was not only a drunkard, but a proselytizing one, plied them with wine all the time they were there. He tried to take Gage on as a chaplain, 'but greater matters called me further, and so I took my leave of him'. Regretfully, the governor let them go to their boats, after giving them 'some dainties of fresh meat, fish and conserves'.

On the morning of the next day, they reached Portobello.

Portobello was at once a very small and a very important town. Its permanent population was barely more than two thousand, yet it was, like Panama, one of the most important commercial centres of the Spanish Empire. Accordingly, it was fortified as if the king himself resided there.

First of all, it was the Caribbean terminus of the treasure route across the Isthmus of Panama. Here came the mule-trains from Panama, laden with bullion and jewels from Peru, which could be loaded at Portobello on to the ships of the Atlantic merchant fleet. Twice in the year, a great fair was held. It was in the form of an open-air bazaar; not only was unimaginable wealth in bullion and precious stones transhipped here, but goods poured in from all of Central America, and it was the great slave market of the area.

The town was protected by three castles, two of them at the mouth of the harbour, and another within. No ship, faced with such cross-fire, could force a passage. The town itself, which was almost two miles within the harbour, was almost an appendage of the citadel of Santiago de la Gloria, and was further protected by a smaller castle, San Jeronimo.

When Gage's small vessel penetrated the harbour, he was sorry to note that there were no great ships riding there. The longer the time before the galleons arrived from Spain, the more money he would have to spend, and he did not have much.

First he had to find lodgings. At that time, they were both plentiful and cheap; indeed, in more than one place he was offered lodgings for nothing, provided only he promised that, when the galleons came, he would either leave them or consent to pay very high rent for them. Over the forty days of the fair, shops might rent for, very roughly, the equivalent of £333 and large houses for £1666.

Luckily, Gage fell in with a 'kind gentleman', who, as it happened, was the King's Treasurer. This man promised that he would find him lodgings which would be cheap even after the fleet had come in. He accompanied Gage in his search, and made use of his authority when they spoke to landlords. They did, at last, find a tiny room: it was just big enough to hold a bed, a table, and a stool or two, 'with room enough besides to open and shut the door'. It rented, for the period of time the fleet was in Portobello, which was about a fortnight, for six-score crowns, about £40. It was, as we have seen, a small town; yet the fair brought a horde of wealthy merchants from Peru, Spain and other places, as well as the four or five thousand soldiers who guarded the fleet. Very often the town was not big enough to contain them all.

Gage himself knew of a merchant who, for fifteen days rental of a shop 'of reasonable bigness' in which to sell his wares, had paid a thousand crowns. But Gage was not a rich merchant, and he did not feel inclined to pay six-score crowns for a room which was nothing more than a mousehole. He threw himself on the mercy of the King's Treasurer, explaining that he had been lately robbed at sea, and could not afford to pay so much for rent alone, since it would leave him no money for food. The landlord, he added, would not cut his price by a farthing. The Treasurer, who does seem to have been an extraordinarily kind man, offered to pay half the rent, if Gage himself could pay the rest. Even this was difficult. Gage was released from his dilemma when another person—he does not say whom—offered him a room for nothing.

He then set out to make a little bit of money for himself, which he did by celebrating a few Masses, and preaching two sermons at fifty crowns apiece.

Gage dutifully visited the castles, but does not tell us anything other than that they were very strong. What did astonish him was the sight of mule-train after mule-train arriving from Panama, loaded down with wedges of silver. In one day, he counted two hundred mules laden with nothing but silver, and watched the wedges of precious metal being unloaded in the public market-place, where they lay 'like heaps of stones in the street, without any fear or suspicion of being lost'.

Within ten days the fleet came: eight galleons and ten merchant ships. Suddenly the streets, which had been quiet and drowsy, were filled with a gesticulating crowd.

Now prices began to climb, steeply and steadily. A fowl, which used to cost one real, now cost twelve; beef, which had cost half a real for thirteen pounds now cost two reals a pound. Even fish and tortoise-meat, the food of the poor, rose in price; and this was all he could afford. He marvelled to see how the merchants sold their commodities, not by the ell or yard, but by the piece or by weight, and to observe that they did not pay in coin, but in wedges of silver which were weighed and bartered like commodities. He was glad to see the wedges of silver disappearing into the bowels of the ships: the faster they were put aboard, the faster the ships would leave, and he with them. Not only was he running out of money; he had a justified fear of falling sick.

Portobello was a fever town at the best of times, and the arrival of so many people put a strain on its limited sanitary facilities. Gage describes it as 'an open grave'. In the year he was there, about five hundred of the merchants, soldiers and marines died of fevers, dysentery, and 'other disorders'. There was a large hospital in the town, with many friars 'whose calling and profession is only to cure and attend upon the sick, and to bear the dead unto their graves'.

The Admiral, Don Carlos de Ybarra (the same who had been in charge of the fleet when Gage had sailed from Cadiz twelve years earlier) was anxious to be gone. Since Gage was afraid of pirates, he thought it would be most prudent to book his passage in one of the biggest and strongest ships, but its captain would not accept him for less than three hundred crowns, which was more than he could afford. Once again, though, his friend the Treasurer came to his rescue. He commended him to the master of a merchant ship, the *San Sebastian*, who was, he knew, anxious to carry a chaplain with him at his own table. The master, in fact, consented to carry him for nothing, and to board him at his own table, only asking in return his prayers for a safe voyage. He offered further, some payment for any sermons Gage should preach in his ship. And Gage 'blessed God, acknowledging in this

also His Providence, who in all occasions furthered my
return to England'.

Thus Gage left America. The year was 1637. He had
avoided, or escaped from, cannibals, sharks, the Inquisition,
caymans, pirates and a fate worse than death in the Philippines. Later, he would return of his own free will to America,
and he would die there.

XXI—*The Apostate*

A DAY or two before Christmas, in the year 1637, Gage finally
landed at Dover. Two days later, he presented himself at the
house of his kinswoman, Lady Penelope Gage. At first she
did not recognize him. This is not strange. He had been
away twenty-three years, having left England as a boy, and
he was so unused to speaking English that he sounded, she
told him, 'like an Indian or a Welshman'.

He had, with his usual bad luck, come home at an unfortunate time. The Great Rebellion was brewing.

All over Europe, new capitalists and old land-holding
aristocrats were rising in revolt against the growing powers
of centralized monarchy. In France, there would be the
Fronde (1649–1653); in the Spanish Empire there would be
the great Catalan revolt of 1640, the revolt of the Portuguese nobles in 1640, the attempt of the Duke of Medina-
Sidonia and the Marquis of Ayamonte, in 1641, to make
Andalusia an independent kingdom, and the Sicilian and
Neopolitan revolts of 1647. Foreign powers took advantage of this chaos, and in the process created a complex web
of international intrigue: Cromwell made overtures to Cardinal de Retz, one of the leaders of the Fronde: Richelieu

helped to arouse the Covenanters against Charles the First, made an attempt to bring the Portuguese rebels under his wing, and put the armies of France on the side of the Catalan republicans; the Jesuits, in addition to preaching, if Voltaire is to be believed, the doctrine of justifiable regicide, supported the Portuguese actively and the Catalans passively; and the Jews financed all comers. The period coincided with a great revival of occultism, from complex hermetic philosophies to the crudities of witchcraft; and the age even had its own international 'philosophes', such as Hartlib, Dury and Comenius.

As in all revolutions, it was hard to tell the reactionaries from the radicals. Representatives of revolutionary capitalism justified themselves by an appeal to ancient privileges; advocates of that new thing in Europe, the monolithic state, fought in the name of loyalty to Church and Crown. Tradition and innovation fought on both sides, and sometimes in the same breast. In the intellectual wars which accompanied the political ones, there appeared the ancestors of almost all contemporary political and social doctrines.

In England, the situation was no less paradoxical and confusing than it was elsewhere. The Great Rebellion has been considered variously a lower-class revolt, a revolt of pragmatism against medieval obscurantism, a recrudescence of medieval millenarism, a capitalist rebellion, a baron's war, a Protestant uprising, and an English expression of an international revolutionary movement. Modern opinions on the matter are as contradictory as those on the French Revolution, and only the uninformed know exactly what happened. Even the actors in this bloody tragedy (or, if you will, glorious epic) stumbled through their parts without a clear idea of what the outcome would be. Cromwell's letters and speeches suggest a choking and sulphurous fog filled with muffled exclamations and the clash of weapons: to the Royalists, the rebellion was an overflowing of the social sewers. It is therefore not surprising that poor Gage, who had come home only to find the house on fire, should have suffered a kind of moral breakdown.

The first blow was the discovery that his father, who had died four years earlier, had indeed cut him out of his will. Gage, though a member of a rich and distinguished family, was penniless and despised in his own country. It is true that he would have protectors so long as he remained a friar. He was in contact with many of the English and Irish Dominicans, Benedictines and Jesuits who clustered around the wealthy English Romanists, and who frequented the court of Queen Henrietta Maria. Once he was arrested, but Sir Francis Windebanke, the Secretary of State, who sympathized with the Roman Catholics, intervened on his behalf. On the other hand, it seems he could not get along with his fellow Dominicans. The Provincial of the small Dominican community was one Thomas Middleton, or Dade. Gage, with two others, Popham and Craft, intrigued to have Middleton removed. Eventually a papal envoy came from Rome to settle the matter. His recommendation, a typically bureaucratic one, was that a new Provincial be appointed, and Gage removed to another area.

This information comes from the letters of George Conn, the Papal agent in England. Gage tells us nothing of the matter. He merely states that he sought and obtained permission to go to Rome, 'to confer some points' with the Master-General of his order. What these points were, we do not know; but one is inclined to be sceptical of Gage's claim that he wished to observe the Papist and Protestant churches of Europe, in order to resolve his doubts. It is more likely that he was still pursuing his 'career' in the Roman church.

He tells us, as it were by-the-by, that he was 'on some business' for his uncle, John Copley of Gatton.

Gage does not tell us what his uncle's business was, but it must have been an important matter. In 1639, we find him on his way to Rome. The governor of Dover Castle, on Gage's presenting him a letter from 'a Papist in London', allowed him to board the packet boat to Dunkirk without being searched. In Flanders, he met his brother Henry, who was stationed with the English Legion near Ghent, in the service of the Spanish crown. He also met his brother's chaplain, the Jesuit Father Peter Wright. This was to prove, for Wright, an unlucky encounter. Henry gave him a letter of recommendation to Don Rodrigo Pacheco Osorio, the Marqués de Serralvo, who was then at Brussels. The Marqués, and 'other great men', gave him letters to many important personages in Rome, among them Cardinals Cucua, Albornos and Barberini.

What was the business which was to bring the least distinguished member of the Gage family, a known runaway and trouble-maker, into the company of some of the greatest personages of Roman Catholic Europe? Francisco Barberini was an enormously powerful figure in Rome. He came from one of the great feudal families of Tuscany. He was the nephew of the Pope (Maffeo Barberini was Urban VIII), and his brothers were Taddeo Barberini, general of the papal forces, and Antonio Barberini, 'Protector of the Crown of France'. Francisco was himself styled 'Protector of the Crown of England'. As to the Marqués de Serralvo, he was that very viceroy who had landed with Gage at Vera Cruz; he had returned to Spain in 1635, much wealthier than he had been when he left. It is true, of course, that Gage's family connections could explain these encounters. It is quite possible that George Gage, who was well-known in Rome, was familiar with 'The Protector of England'; and Sir Henry and the Marqués de Serralvo were both in the service of the Spanish crown in the Low Countries. Yet one cannot help wishing that Gage had stated exactly what his uncle John's 'business' was.

There was much talk in Rome, Gage tells us, about Archbishop Laud. Laud, he would have us believe, was secretly a Roman Catholic. He must have known better. It is true that many Roman Catholics thought, at one time, that Laud might be won over: indeed, the Pope, as Laud himself recorded in his diary, had tempted him with the offer of a cardinal's hat. But it soon became clear to Rome that Laud was as Anglican as his king, and that the peace he desired with Rome was the peace of equals. Such confidence was more galling than the fury of the most rabid Calvinist; and Rome came to the conclusion that Laud meant, to use the words of a contemporary Roman Catholic Englishman, William Weston, 'to frame a motley religion of his own, and be lord of it himself'. Gage's statements about Laud are, in fact, Puritan propaganda, intended to justify his martyrdom as an act of justice.

He also passes along some gossip about a few noted recusants who were, he says, being considered for cardinalates. These were, he says, Sir William Hamilton, who was backed by 'a great number of friends', Sir Kenelm Digby, who was sponsored by Queen Henrietta Maria and assiduously puffed by his chaplain, Fitton, and two other prominent English papists—Walter Montague, son of the Earl of Manchester, and Sir Toby Matthew, a great friend of his brother George. Sir William Hamilton, however, had become involved with a woman, and was now out of the running.

This, while it may be factually true, is misleading. Gage, in his anxiety to condemn the English at Rome, does not tell us that Montague, Digby and Fitton were regarded with some suspicion by the Vatican. They were far from being orthodox. Digby was an occultist, and perhaps an early Freemason (in a paper read before the Royal Society in 1661, he would refer to God as the 'Supreme Architect'), and he used to hold seances with Walter Montague. Freemasonry, certainly, had not yet been condemned by Rome; but the trio also had well marked tendencies towards theological heresy, as Rome then defined it. In 1646, they would become open 'Blackloists', adherents to the programme of Thomas White, who wrote under the pseudonym of 'Blackloe', among others. White was a teacher of philosophy at Douai College, which, curiously enough, would later be headed by Thomas' half-brother Francis. White wanted a National English Catholic church free of Vatican control; he questioned the infallibility of the Pope, and declared that he should be obeyed only when he demanded nothing contrary to civil loyalties; he stated that Catholics should be at least passively obedient to all established secular power; he opposed the Inquisition; and he was strongly anti-Jesuit. 'Blackloism' was, for obvious reasons, unacceptable to Rome, and White's books were put on the Index. It was said by his enemies that White's 'English Jansenism' was simply a device to induce Cromwell to accept an English Catholic church based on such principles. Indeed, Digby later became Cromwell's confidant, and his go-between with Cardinal Mazarin.

One would be surprised if Gage was unaware of Rome's suspicions (he tells us that the candidates failed); but he was obviously confident that his readers would not know them. He did not lie to them; rather, he retailed truths in such a way as to imply falsehoods.

Gage surfaced, from this immersion in the deep and sombre pool of Roman intrigue, to return to England. He had already made up his mind, he says, to become an Anglican.

In 1642, he preached a sermon of recantation in St Paul's, in London. By 1648, when his book was published, he had 'resolved upon a choice for the Parliament cause'.

In 1642, Father Thomas Holland was brought to trial as a Roman Catholic priest who had said mass in England. It was Gage's testimony which convicted him: he had been a schoolmate of his at St Omer's. Father Holland was hanged, drawn and quartered at Tyburn, praying to the end for Gage's soul.

Frightened by the appearance of a Protestant fanatic in the midst of his family, Colonel Henry Gage promised his brother a thousand pounds if he would leave England, but Gage was not to be tempted, at this point, by money.

In 1643, Gage's testimony was responsible for the execution of Father Arthur Bell, a Franciscan.

During the next two years he worked on *The English-American, or A New Survey of the West Indies*, and in 1648 it was published.

The book was an immediate success. It contained an advertisement, in awkward and confused but at times oddly imaginative verse, by Thomas Chaloner, a Yorkshire businessman and adherent of 'the natural religion', who was powerful in the Commonwealth, and was to become infamous as one of the regicides. Furthermore, Gage was now. rector of Acrise, a small parish in Kent, and he had taken a wife 'to satisfy the world of my sincerity'. Things were going well for him.

In his dedication to Lord Fairfax, 'Captain-General of the Parliament's Army', Gage sets out the reasons why he thinks his book might be 'to the use and benefit of my English countrymen'. Its main purpose, we discover, is to serve as a guidebook for an invading army.

He then discusses the reasons which might justify an English invasion of Spanish America.

These are well-considered and concisely expressed, though completely unoriginal. Such an enterprise would occupy the men of Cromwell's army after 'The settlement of the peace of this kingdom, and reduction of Ireland', and would ensure that their idleness 'may neither be a burden to themselves not the kingdom'. Englishmen, having established plantations in Barbados and other Caribbean islands, had become adapted to the climate, and were, in any case, more than halfway there. The conquest would be easy; witness the success of the Dutch against the Spaniards.

He then examines the objection, 'that the Spaniard being entitled to those countries, it were both unlawful and against all conscience, to dispossess him thereof'. Against this, he sets an argument with which we are all familiar: he disguises imperialism as anti-imperialism. 'No question but the just right or title to those countries appertains to the natives themselves, who, if they shall willingly and freely invite the English to their protection, what title soever they have in them no doubt they may legally transfer it or communicate it to others.' The Spaniards have no rights of ownership, but those based, first, on 'discovery', which in no way justifies the seizure of the 'discovered' lands from those who truly own it, and second, on force, which must yield to greater force. Furthermore, the atrocities committed by the Spaniards have invalidated their claim that they have a right to the possession of the New World, because of the moral superiority of their civilization over that of the Aztecs and other Indian peoples. Finally, as to the land unoccupied by either Spaniards or Indians, England has as much right to it as does any other power.

We may recognize here, briefly expressed, some of the

leading ideas of the imperialist party, ideas which had altered little from Elizabethan times. The hoped-for conquest of Spanish America, either by military or by economic means, was to be an important factor in British foreign policy to the days of Pitt. At the end of the sixteenth and beginning of the seventeenth centuries, its chief intellectual spokesmen were Raleigh, the Hakluyts and Bacon, who were supported by a horde of pamphleteers. Most of its practical support came from merchants interested in trans-Atlantic trade, a group particularly concentrated in the Virginia and Providence Island Companies. It was in these circles, also, that revolutionary feeling was concentrated. Christopher Hill, the historian, has described the Providence Island Company as 'a cover-organization for the Parliamentary opposition to Charles'. Three other groups were also interested in the conquest of Spanish America—the New England merchants, who had connections with the above-mentioned group, the planters of Barbados, and the Marranos, who had substantial interests in Barbados, the maritime trade of New England, and Spanish America. Cromwell himself had been a close associate of the founders of the Providence Island Company, and had been a member of the parliamentary commission for Plantations.

Before Gage was to receive the patronage of such exalted circles, however, he was to commit one act which has made his name infamous.

In 1651, there took place the trials of Fathers Middleton and Peter Wright. The last-named was Sir Henry's chaplain, and the man who had comforted him as he lay dying for his king near Oxford: Sir Henry had come over to England to fight for Charles. Gage was commanded to come up to London to give evidence.

He argued that while Father Middleton was a friar, he

was not a priest, and as a result of this testimony Father Middleton was acquitted. When he testified against Father Wright, though, he behaved in an altogether different manner, at least according to the report which has come down to us. One witness describes him as raving for nearly thirty minutes 'in so violent and marked a manner that it was evident from the beginning that he was instigated by hatred and a thirst for blood'. His own brother, he said, had planned his murder, and Father Wright, though he had heard of the plan in the confessional, had done nothing to prevent it. Father Wright declared he knew nothing of such a plan; nevertheless, he was executed at Tyburn on the tenth of May, 1651.

George Gage had already done his best to dissuade his brother from giving such damaging testimony. George had gone, says a Jesuit historian, 'to the haunt of vice where the wretched debauchee was lodging . . . to warn him of the divine judgments that were hanging over him if he should make himself guilty of innocent blood'. This 'haunt of vice' was, presumably, Gage's home.

In 1652, the Reverend George Gage was himself put in prison, where he later died. In a bill against him for appearance before the courts, Thomas is cited as an informer against him. There is no evidence that Thomas was heard in court, and Gage's editor, Thompson, has suggested that the bill might have been referring to information gleaned from his book. However, it must be admitted as unlikely that, if information was gleaned in this roundabout manner, a court document would not make some reference to the fact.

Gage's behaviour at this stage in his life is most puzzling. So far we have seen him to be hypocritical in his religion, and venal and unscrupulous in money matters; on the other hand he has shown that he could be generous to those dependent on him, and that he was not uncourageous. We have even seen him to be, if only occasionally, a man of conscience.

There is no doubt that Gage felt he was testifying against the instruments of a criminal conspiracy. There is no doubt also that his new Puritan friends considered him, not a traitor, but 'a brand snatched from the burning'. Certainly the aims of the Roman Catholic activists of the time were all too clearly treasonable, and were recognized as such by most Englishmen, whether they were Puritans or not. Gage felt that his testimony put him in danger of death; and, indeed, he asked the court to protect him against attempts on his life, a request which was not granted. As to his 'thirst for blood', most of the accounts of his behaviour at the trials come from Jesuit and Dominican historians, from whom objectivity can hardly be expected. Gage cannot, after all, have been actuated entirely by malice, since he saved the life of Father Middleton, with whom he had quarrelled. He may also be forgiven if his sense of family solidarity had been somewhat weakened. Not only had his father cut him off without a penny, but his brother had tried to have him killed.

'What I have been able to discover for the good of this state I have done, and not spared (when called upon) to give in true evidence upon my oath against Jesuits, priests and friars; for the which (after a fair invitation from my brother Colonel Gage to come over again to Flanders, offering me a thousand pound ready money) I have been once assaulted in Aldersgate Street; and another time like to be killed in Shoe Lane by a captain of my brother's regiment, named Vincent Burton, who (as I was after informed) came from Flanders on purpose to make me away or convey me over, and with such a malicious design followed me to my lodging, lifting up the latch and opening the door (as he had seen me do) and attempting to go up the stairs to my chamber without any enquiry for me, or knocking at the door; from which God graciously delivered me by the weak means of a woman my landlady, who stopped him from going any further; and being demanded his name, and answering by the name of Stewart, and my landlady telling him from me that I knew him not, he went away chafing and saying that I should know him before he had done with me. But he that knoweth God well shall know no enemy to his hurt; neither have I never since seen or known this man. I might here also write down the contents of a threatening letter from mine own brother, when he was colonel for the King of England and governor of Oxford, but forbear to do so out of some tender considerations of flesh and blood . . .'

This said, however, his actions at this time remain inexplicable. The most favourable adjective one could apply to them, even considering all the pressures to which he was subject, would be 'contemptible'; and perhaps a much stronger adjective would be more appropriate.

In 1654, Cromwell, having reduced the Irish and almost everything and everyone else ('The Lord is pleased still to vouchsafe us His presence, and to prosper His own work in our hands'), began to dream of Empire. As he was to say two years later, in one of those curiously murky speeches of his:

'The Spaniard is your enemy; and your enemy, as I tell you, naturally, by that antipathy which is in him—providentially, and this in divers respects. You could not get an honest or honourable peace from him; it was sought by the Long Parliament; it was not attained. It could not be attained with honour and honesty. I say, it could not be attained with honour and honesty. And truly, when I say that, he is naturally throughout an enemy; an enmity is put into him by God. "I will put an enmity between thy seed and her seed"; which goes but for little among statesmen, but is more considerable than all things! And he that considers not such natural enmity

189

the providential enmity, as well as the accidental, I think he is not well acquainted with Scripture and the things of God. And the Spaniard is not only our enemy accidentally, but he is providentially so, God having in His wisdom disposed it so to be, when we made a breach with the Spanish nation.'

Gage, perhaps on Chaloner's recommendation, was asked to prepare a memorandum on the prospects of success of an invasion of Spanish America. He did so; and the gist of it was that conquest would be very easy. All the arguments were in his book. We have seen the main reasons for Cromwell's interest in this scheme; but a further reason is given by Bishop Burnet, in his *History of His Own Times*: 'By this he reckoned he would be supplied with such a treasure, that his government would be established before he should need to have any recourse to a parliament for money.'

The Venetian ambassador in London informed his government that plans were being laid for an attack, based on 'the information of a Dominican friar who has been in those parts, and knows them well, who has had many secret conferences with the Protector on the subject'. And one Stoupe, coming in upon Cromwell one day, saw him examining, with loving attention, a chart of the Gulf of Mexico.

Milton, who was of course a civil servant at the time, was to write a tract, in which he contended:

'that the motives whereby we have been lately induced to make an attack upon certain islands in the West Indies which have been now for some time in the hands of the Spaniards are exceeding just and reasonable, every one will easily see who considers in what a hostile manner that king and his subjects have all alone, in those parts of America, treated the English nation . . .'

In December of 1654, the invasion fleet left Portsmouth, under sealed orders. A declaration of war would not follow until December 1655; Cromwell did not believe in giving his enemies the advantage of a warning. (The advantage of fighting with the forces of Anti-Christ, as compared to those of fighting with only relatively evil men, is that one is not compelled to observe the niceties of war.) There were sixty ships, and four thousand soldiers, which last number was augmented, at Barbados, to a number estimated by various historians as being from 6,680 to 9,000.

Gage was chaplain to General Robert Venables, who was, with Admiral William Penn, co-commander of the expedition. Their objective was Hispaniola.

The attempt to invade Hispaniola was, however, a disaster.

To begin with, the soldiers were on the verge of mutiny. They were not to be allowed any booty: gold and plate was to be turned over to their officers, to be put away for the Lord Protector. Their reward was to be miserly in the extreme: six weeks extra pay.

The officers, by a combination of threats and promises, finally got the troops ashore. They came across an Irishman, and asked him to lead them to a spring, where they might fill their water-casks. But he led them straight to a Spanish fort, and when the Spaniards opened fire, General Venables hid behind a tree.

Later encounters were equally disastrous.

No doubt Venables blamed Gage for the complete failure of the invasion. Gage had argued that the negroes of Hispaniola would immediately ally themselves with the English against their Spanish masters. In fact, however, the negroes had proved loyal: their masters had promised them their freedom if they fought, and had persuaded them, moreover, that the English were cannibals. As for the Spaniards, they fought with supreme confidence, since they had indulgences hanging about their necks, pardoning them from all sins past and to come.

The British had withdrawn and camped on the beach. There were many wild cattle roaming in the forest, but they were afraid to enter it. After a while, since they were on the verge of starvation, they began to eat their own dogs and horses. At night the land crabs came down to the sea to feed, and the British, hearing the clicking of their legs and thinking them the rattling of bandoliers, ran into the sea. They shot at fire-flies, thinking them the lights of the enemy.

When they began to come down with various tropical diseases, they put it to General Venables that they should leave Hispaniola and try Jamaica instead. They left behind them a few Spanish dead, and a desecrated image of the Virgin Mary, which they had pelted with oranges.

The behaviour of this pathetic rabble need not surprise us. Consider who composed it. The two thousand five hundred Englishmen were mostly the dregs of the Parliamentary army, or vagabonds who had been recruited by force and given no military training whatsoever. With them were an undetermined number of beachcombers, failed planters, and unsatisfactory servants from Barbados and other West Indian islands.

Not all of them were of British stock, however. The pilot, Campoe Sabbatha (or Sibada) was a Marrano from Antigua, and another Marrano, Acosta, was in charge of the commissariat.

There were also some proto-Americans. For example, Robert Sedgwick, the merchant from Charleston, was a Major-General in the expedition.

Had Gage stood upon the deck of the flagship as the fleet entered what is now Kingston Harbour, he would have seen no sign of a city. Where the port is now, he would have seen only bush, and the mangroves growing into the shallow water, a wave-lapped thicket of roots. The only sign of human occupation was Passage Fort, an unimpressive fortification.

The *Martin*, with twelve guns and a crew of sixty, was one of the smallest ships in the fleet. But, of the few shots it fired into the fort, one disabled a Spanish major, the best soldier there. The *Martin* was run ashore, and its men disembarked. The Spanish defenders fled, leaving three guns mounted in the fort. In the taking of Passage Fort, not one English life was lost.

When the soldiers had landed, they marched through a wood until they came to the edge of the savannah. Their prize, Santiago de la Vega, was only a quarter of a mile away. It was an attractive town of some nine hundred houses. The houses were simple enough. They were rambling single-storey structures, made of the split trunks of cabbage palms, or of reed overcast with mortar and lime, and roofed with tiles or palmetto thatch. There were six notable structures —five churches and a Franciscan cloister. Santiago de la Vega was laid out in the grid-pattern typical of Spanish colonial towns, and the houses were, except for those around the market-place, set far apart. It was deserted: the Spaniards had buried their treasures, turned their livestock loose, and fled to the north.

The British sacked and burned the town, and melted down the church bells to make bullets.

What were Gage's feelings when he saw the destruction of Santiago de la Vega? Did he have qualms of conscience? Did he think pleasantly of the land grant which would come to him as one of the pioneers in the conquest of Jamaica? Was he proud or ashamed of his role in the action; and did he still feel, as he had said in his book, that God had made use of him 'as a Joseph to discover the treasures of Egypt'? How had he felt when, in Hispaniola, the soldiers had splattered with oranges the statue of the Virgin Mary, with the infant Lord in her arms? Had he remembered, perhaps, a more innocent game in a little town near Chiapa de Corzo, when he and his friends had thrown oranges at the sub-deacon Juan Escudero?

We do not know. All we know is that Thomas Gage died in Jamaica early in 1656. The island was still ruled by military law, and there were no other clergymen, since the other chaplain, William Dennys, had died already. Gage's death was, in all probability, not an easy one.

In November, 1655, a few months before Gage died, a member of the expedition, whose name has now been lost, wrote the following letter home—

'The eleven ships lately arrived at this place with &c. poor men I pity them at the heart, all their imaginary mountains of gold are turned into dross, and their reason and affections are ready to bid them sail home again already. For my part, greater disappointments I never met with, having had no provision allowed me in ten weeks last past, or above three biscuits this fourteen weeks, so that all I can rake and scrape in ready money goes to housekeeping, and the shifts I make are not to be written here. We have lost half our army from our first landing in Spaniola, when we were eight thousand, besides one thousand or more seamen in arms. Never did my eyes see such a sickly time, nor so many funeral and graves all the town over that it is a very Golgotha. We have a savannah or plain near us where some of the soldiers are buried so shallow that the Spanish dogs, which lurk about the town, scrape them up and eat them. As for English dogs, they are most eaten by our soldiery; not one walks the streets that is not shot at, unless well befriended and respected. We have not only eaten all the cattle within twelve miles of the place, but now almost all the horses, asses, mules' flesh near us, so that I shall hold little Eastcheap in more esteem than the whole Indies if this trade last, and I can give nor learn no reason that it should not here continue; so beside this we expect no pay here, nor hardly at home

194

now, but perhaps some ragged land at the best, and that but by the by spoken of, for us general officers not a word mentioned. I could dwell long upon this subject, and could tell you that still half our army lies sick and helpless, nor had we victuals for them before this fleet, nor expect aught now save some bread, and bran, and oatmeal, and if that with physic will not keep them alive, we have no other remedy but death for them. For my own part in twenty-five years have I not endured so much sickness as here with bloody flux, rheum, ague, fever, so that I desire earnestly to go for England in March next, if permitted, for I am fallen away five inches about.'

It was in surroundings such as this that Gage died. We do not know the cause of his death, but it might have been one of the diseases which were then sweeping through the army of invasion. He did not even have the satisfaction of knowing that, when the news of Jamaica's fall reached Mexico, all the church bells of the viceregal capital tolled in mourning.

It was, if belated and unsatisfactory, some kind of recognition.

Appendix

Was Gage a Spy?

CERTAIN inconsistencies in Gage's story suggest ambiguities in his motives in going to Mexico. I may be unduly suspicious in noting these, but at least they occasion interesting speculations.

It must be admitted, at the beginning, that it is unlikely that the questions I raise here will ever be solved. The fact is that the little evidence we have concerning Gage's early life is contradictory. We do not know the exact year of his birth; and indeed we cannot even be sure of his place of birth: some sources say he was born in Ireland. In his deposition against Thomas Holland, Gage stated that he and Holland had studied together for five years at St Omer's, a statement Holland did not deny. This would mean that Gage was at St Omer's between 1615 and 1621, as Holland is known to have been. However, there is other evidence which seems to indicate that Gage was studying theology in Spain in 1615. (In his book, *A Hundred Homeless Years*, Father Godfrey Anstruther, O.P., goes into the chronology of Gage's early life at some length.)

We need not read, into this confusion of dates, a deliberate attempt at obfuscation. It simply indicates the unreliability of much apparently 'hard' documentary evidence. To illustrate this point, let us take the evidence relating to the arrival in Mexico of the Marqués de Serralvo, who, Gage says, landed with him in Vera Cruz on the twelfth of September,

1625. One respectable Mexican historian says Serralvo arrived on the third of November, 1624; one of the best United States authorities on colonial Mexican history has him arriving in November of 1626; another U.S. authority in this field seems to believe he was well established in Mexico in 1624. When there is so much confusion concerning an important event in Mexican history, we have no right to expect complete certainty in the biography of an obscure Dominican friar.

We cannot be sure, therefore, of the dates we have. Least of all can we be sure of Gage's exact age. This should be kept in mind when considering the speculation that follows.

Thompson, in his introduction to his edition of the *Travels*, points out that the records of the Privy Council for 1616–17 state that a certain Thomas Gage had been arrested in 1617, held incommunicado as 'a dangerous man', and brought before the Privy Council. A. P. Newton believes this was our Thomas Gage, that he had come back to England to see his family, and that he was arrested because it was known he was returning to St Omer. 'After a few days detention', Newton says, 'he was released, and went back to his studies.' Thompson believes this could not have been Gage, who was, he says, only fourteen or fifteen at the time, and thus hardly a 'man': in any case, he adds, students at St Omer did not return to England until their studies were completed. He thinks that somebody else may have assumed the name 'Gage', since it was common for travelling recusants and priests to adopt non-existent names, or to exchange names, to confuse the authorities.

This is certainly logical. However, the assumption that Gage was only in his mid-teens at the time is based on deductions and probabilities, rather than a sure knowledge of the facts: it is equally possible that he was in his late teens

197

in 1617, since his exact birth-date is not known. Further-more, if Gage was only in his mid-teens in 1617, then he would have been too young in 1625 to have become a Jesuit. Thus his father's action in 1625—cutting him out of his will for antagonism to the Jesuits—would have seemed excessive, as Thompson himself says. Thompson suggests that Thomas had given out 'recriminations and perhaps sensational charges against the order which would be given added emphasis by enemies of the Papacy because they had been uttered by a member of one of the prominent families of England that had stood by the old faith and suffered for it'. But the action would still have been excessive, if Gage had simply slandered the order in conversation, particularly if he was not a member of the order, and was still a boy.

Furthermore, Gage's fears that he would be completely scorned by his friends and kindred if he returned to England would seem a baseless hysteria, if he had only refused to become a Jesuit, even if this was accompanied by boyish complaints against the order. The rivalry between religious orders in the seventeenth century is well known: one did not cease to become a good Roman Catholic by hating the Jesuits.

One cannot help suspecting that Gage had, in some manner, done serious harm to the Roman Catholic cause.

Christopher Devlin's life of Robert Southwell, though it deals with the reign of Elizabeth, an earlier time, is en-lightening on the methods used by English agents to frustrate the work of the Jesuits.

For one thing, they always approached monks and friars who had differences with their superiors: Gage had a notably rebel-lious and factious spirit, as research into his life clearly shows.

Devlin believes that the playwright, Gosson, was rewarded for his spying by being made rector in the Parish Church at

Stepney. Gage, after he left the Roman Church, was given the church of Acrise, in Kent.

Could the 'dangerous man' taken in 1617 really have been our Thomas Gage, and could he then have been encouraged to act as a counter-intelligence agent for Protestant England? Could his father have heard of this in some way? If this were true (and I must emphasize that this is only a wild hypothesis), it would account for his black rage against his son, which would seem excessive in view of the facts as Gage presents them. Certainly, it was not uncommon for the government to capture suspected priests, and release them on assurance that they would provide them with information.

My suspicion was first aroused by Gage's extraordinarily vivid and careful descriptions. The book was published in 1648, eleven years after he left America. I find it difficult to believe that, unless he had a photographic memory, he could have remembered what he did. He must have taken notes as he went, since there were few books of Mexican geography to which he could return to refresh his memory. Why did he take notes?

Note-takers are not necessarily spies, one might say, and this is perfectly true. To this one might respond, though, that he seems to have made the kind of notes one would expect a spy to make. He makes a careful analysis of the economic state of the country: he notes major roads and navigable rivers, gives an estimate of the population of various cities, towns and villages, and is precise on the details of fortifications. If all travelling friars had had his strategic sense, the entire world would have been Roman Catholic long ago.

The closing paragraph of his book is significant—

'I shall think my time and pen happily employed if by

what here I have written I may strengthen the perusers of this small volume against Popish superstition whether in England, other parts of Europe, Asia or America; for the which I shall offer up my daily prayers unto Him, Who (as I may well say) miraculously brought me from America to England, and hath made use of me as a Joseph to discover the treasures of Egypt, or as the spies to search into the land of Canaan, even the God of all nations, to Whom be ascribed by me and all true and faithful believers, Glory, Power, Majesty and Mercy for evermore. Amen.'

Having said all this, I must say that my own opinion is that he was certainly not a spy in the sense that he was secretly in the pay of his government from the beginning. Admittedly, the matter could not be decided until we know what went on in that Privy Council meeting in 1617. But his long stay in the unimportant villages of Mixco and Pinola does not suggest that he was gathering intelligence on his government's orders.

I am inclined to think that he was, as a spy, self-employed. I feel that he became disillusioned with Roman Catholicism quite early, that his stay in America completed the process, and that he was not unaware of the possibility that, if the worst came to the worst, his notes could be put to use if he left the Church. I would say that in Guatemala he actively began thinking that his notes might be of use to the British Government, since it is at this point that he begins noticing fortifications and possible invasion routes, not with the idle speculations of a traveller, but in a more conscious and detailed way. It is in Guatemala that 'four-eyed Hidalgo' (he may not merely be referring to his glasses) suggests that he might be a spy and a heretic, and it is here that he makes the acquaintance of Antonio Fernandez, the 'Portuguese' merchant, who departs from Guatemala 'for some reasons which here I must conceal'. Here also he begins to show some knowledge of Lutheran and Calvinist doctrine, though it must be said that he could have acquired some knowledge of these doctrines by reading Papist arguments against them. In his book *The Fall of the Spanish-American Empire*, Salvador de Madariaga has said (with adequate supporting documentation), 'The Jews were assiduous agents in the dissemination of the Reformation, less out of any genuine interest in the Reformation as such than because it implied a schism and a division in the rival faith' (p. 248).

Furthermore, as is well known, the Marranos in both America and Europe, hating the Roman Catholicism of the Inquisition, did all they could to advance a possible invasion of the Spanish New World by Cromwellian England. Cromwell's Jewish intelligencers—the merchants Carvajal, de Caceres and Dormido—gave him access to extensive and speedy information about the affairs of the New World. Before Cromwell came on the scene, the 'Portuguese' of the New World, and the great Marrano mercantile houses of Europe had been conspiring with the Dutch to the same end.

In other words, Gage's 'Portuguese' friends in Guatemala may have helped him to become a Protestant, and may themselves have engaged in undercover activities.

Rydjord (see Bibliography) has pointed out the role played by the Portuguese in the fomenting of sedition in New Spain, particularly in the 1640's, when Portugal was fighting for her independence. The Inquisition, indeed, imprisoned many Portuguese at this time, so many that the prisons could not hold them all.

The viceroy at this time was Don Diego Lopéz Pacheco Cabrera y Bobadilla, Duque de Escalona and Marqués de Villena; and there were stories that he had talked of making himself king of New Spain. He was, it was well known, a cousin of the Duke of Braganza, who had seized the throne of Portugal. He had distributed pamphlets, pointing out the relationship of the House of Pacheco to the royal families of England, Savoy, Modena and Spain itself. His enemies grumbled at his garrisoning the country's most strategic port, San Juan de Ulloa, with Portuguese soldiers; and, by a curious coincidence, a shipment of treasure from the viceroy arrived in Portugal just as that country rose in revolt. Finally, his bishop, Palafox, suspected him of wanting to set up an independent dynasty in New Spain. In 1643, the Duke of Escalona was removed from his post without warning, and sent to Spain to explain his actions. Philip the Fourth, fully satisfied with his explanation, named him the new viceroy of Sicily, which revolted in 1647.

One would love to know the full title of the 'Duke of Medina', who was aware of the fact that this obscure English friar was sailing to America, and tried, if ineffectually, to prevent his departure. Unfortunately, there were three dukes of this title prominent in the public life of the time—Medina de Rioseco, Medina-Sidonia and Medina de las Torres— and we do not know which 'Medina' Gage is referring to, though the most likely candidate is Medina-Sidonia. Medina-Sidonia, brother of the Queen of Portugal, later hoped— with the help of the Portuguese rebels—to make himself king of an independent Andalusia. Curiously enough, though the island of Jamaica was held in fief by the Dukes of Veragua (the descendants of Columbus, a title which died out, in the direct male line, with Diego Columbus), Henry Whistler, in his *Journal of the West Indian Expedition 1654–55*, says Jamaica belongs to 'the Duke of Meden' (Williams—p. 287).

However, this is all speculation: the facts here gathered can be explained, individually at least. Only an expert in the diplomatic history of the Cromwellian period could properly evaluate them. Even if it hangs together as an argument, and helps to explain certain inconsistencies in Gage's narrative, it by no means follows that the tidiest explanation is the true one. Life and art are two quite separate things; real people are seldom consistent, and attempts to explain loose ends in their behaviour can often lead, I readily admit, to a picture which may pervert the untidy reality into a neat work of the rational imagination.

Acknowledgements and Select Bibliography

I am indebted to Father Godfrey Anstruther, O.P., author of *A Hundred Homeless Years*, for some of the quotations from Dominican documents concerning Gage.

EDITIONS OF GAGE'S TRAVELS

I am greatly indebted to two excellent editions of Gage's travels. One of these, *Thomas Gage's Travels in the New World*, was prepared by the noted archaeologist, J. Eric S. Thompson, and published by the University of Oklahoma Press in 1958. This edition is full of new and interesting material on Gage, much more than I have been able to incorporate in this brief book. The other, *Thomas Gage: The English-American*, was prepared by A. P. Newton, and published by Routledge in London in 1928, in the *Broadway Travellers* series. It is from this that my quotations from Gage himself are taken, with the kind permission of the publishers. Though Thompson's research has invalidated one or two of Newton's points, his edition is still valuable, since it presents some material for which Thompson had no room, and is enlivened by an excellent introduction and some charming prints. Newton's edition is, on the whole, more faithful to Gage's text: Thompson has, for the sake of readability, cut up Gage's exasperatingly long sentences, and rationalized his often confusing grammar.

GENERAL

AKRIGG, G. P. V.: *Jacobean Pageant, The Court of King James I*: Harvard University Press, Cambridge, Mass., 1962

ALAMÁN, Lucas: *Obras*: Editorial Jus, Mexico D.F., 1942

ANONYMOUS: *Annals of the Cakchiquels* (trans. Recinos and Goetz): University of Oklahoma Press, Norman, Oklahoma, 1953

ANONYMOUS: *Popol Vuh* (trans. Recinos, Goetz and Morley): William Hodge and Co. Ltd, London, 1951

ANSTRUTHER, Father Godfrey, O. P.: *A Hundred Homeless Years*: Blackfriar's Press, London, 1959

BALBUENA, Bernardo de: *Grandeza de Mexico (La Grandeza Mexicana)*: Imprenta Universitaria, Mexico D.F.

ACKNOWLEDGEMENTS AND SELECT BIBLIOGRAPHY

BARBOUR, Violet: *Capitalism in Amsterdam in the Seventeenth Century*: University of Michigan Press, 1963

BOURNE, Edward Gaylord: *Spain in America, 1450–1580*: Barnes and Noble Inc., New York, 1962

BURNS, Sir Alan: *History of the British West Indies*: George Allen and Unwin Ltd, London, 1954

CROMWELL, Oliver: *Letters and Speeches* (ed. Carlyle): Routledge and Sons, London, 1888

CUNDALL, Frank: *Historic Jamaica*: Published for the Institute of Jamaica by the West India Committee: London, 1915

DAVIES, R. Trevor: *Spain in Decline, 1621–1700*: Macmillan and Co. Ltd, London, 1957

DEVLIN, Christopher: *The Life of Robert Southwell*: Longmans, London, 1956

DIAZ DEL CASTILLO, Bernal: *Historia Verdadera de la Conquista de la Nueva España*: Espasa-Calpe Argentina S.A., Mexico D.F., 1965

GALINDO Y VILLA: Jesus: *Historia Sumaria de la Ciudad de Mexico*: Editora Nacional S.A., Mexico D.F., 1955

GERARD, John: *John Gerard, The Autobiography of an Elizabethan*: (trans. Philip Caraman): Longmans, London, 1951

HARING, C. H.: *The Spanish Empire in America*: Harcourt Brace and World Inc., New York, 1952

HILL, Christopher: *The Century of Revolution*: Nelson and Sons Ltd, Edinburgh, 1962

HILL, Christopher: *Intellectual Origins of the English Revolution*: Oxford University Press, 1965

Historia Documental de Mexico: Universidad Nacional Autonoma de Mexico, Mexico, D.F., 1964

JESSE, John Heneage Jesse: *London, Her Crowned and Uncrowned, from Westminster to the Tower*: Nicholls and Co, London, undated

LEONARD, Irving A.: *Baroque Times in old Mexico*: University of Michigan Press, 1959

MADARIAGA, Salvador de: *The Fall of the Spanish-American Empire:* Hollis and Carter, London, 1947

MAYER, William: *Early Travellers in Mexico*: Mexico D.F., 1961 (privately printed)

PETERSON, R. T.: *Sir Kenelm Digby:* Jonathan Cape, London, 1956

RICARD, Robert: *The Spiritual Conquest of Mexico* (trans. Simpson): University of California Press, Berkeley, California, 1966

ROTH, Cecil: *A History of the Marranos*: Harper and Row, New York, 1966

RYDJORD, John: *Foreign Interest in the Independence of New Spain*: Duke University Press, Durham, North Carolina, 1935

SCHOLES, France V. and ROYS, Ralph L.: *Fray Diego de Landa and The Problem of Idolatry in Yucatan*: Carnegie Institute of Washington, 1938

SCHURZ, William Lytle: *The Manilla Galleon*: Dutton, New York, 1939

SIMPSON, Lesley Byrd: *Many Mexicos*: University of California Press, 1966

TREVELYAN, G. M.: *England Under the Stuarts*: Penguin Books, Harmondsworth, Middlesex, 1960

WILLIAMS, Eric (Ed.): *Documents of West Indian History 1492–1655*: PNM Publishing Co. Ltd, Port-of-Spain, Trinidad, 1963

ZAMACOIS, Niceto de: *Historia de Mexico* (Vol. 5): J. F. Parres y Cia: Mexico D.F., 1878

Index